W9-AGK-414

The Essential Agrarian Reader

E. 35
CEDAR
ONE WAY

The Essential Agrarian Reader

The Future of Culture, Community, and the Land

EDITED BY
NORMAN WIRZBA

THE UNIVERSITY PRESS OF KENTUCKY

Publication of this volume was made possible in part by a grant from the National Endowment for the Humanities.

Scholarly publisher for the Commonwealth,
serving Bellarmine University, Berea College, Centre College of Kentucky, Eastern Kentucky University, The Filson Historical Society, Georgetown College, Kentucky Historical Society, Kentucky State University, Morehead State University, Murray State University, Northern Kentucky University, Transylvania University, University of Kentucky, University of Louisville, and Western Kentucky University.

Editorial and Sales Offices: The University Press of Kentucky
663 South Limestone Street, Lexington, Kentucky 40508-4008

Photography: Cover: *Terraced Plowing, Saline County, Kansas, September 1990*, © Terry Evans; Frontispiece and page 99 © Gregory Conniff; pages 21 and 189, © Nina Danforth.

07 06 05 04 03   5 4 3 2 1

Library of Congress Cataloging-in-Publication Data

The essential agrarian reader : the future of culture, community, and the land / edited by Norman Wirzba.
p. cm.
Includes bibliographical references and index.
ISBN 0-8131-2285-6 (Hardcover : alk. paper)
1. Agriculture—Economic aspects. 2. Agriculture—Environmental aspects. 3. Agriculture—Social aspects. 4. Human ecology. 5. Agriculture—Moral and ethical aspects. I. Wirzba, Norman.
HD1433.E87 2003
338.1—dc21   2003008808

This book is printed on acid-free recycled paper meeting the requirements of the American National Standard for Permanence in Paper for Printed Library Materials.

Manufactured in the United States of America.

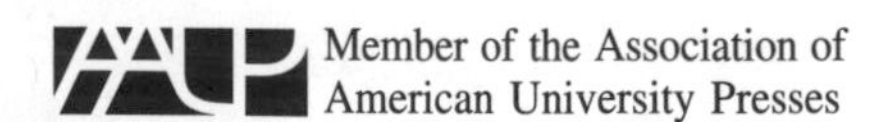

For Wendell Berry
mentor and friend

Contents

Part 3: Putting Agrarianism to Work

Foreword

Barbara Kingsolver

Sometime around my fortieth birthday I began an earnest study of agriculture. I worked quietly on this project, speaking of my new interest to almost no one because of what they might think. Specifically, they might think I was out of my mind.

Why? Because at this moment in history it's considered smart to get *out* of agriculture. And because I was already embarked on a career as a writer, doing work that many people might consider intellectual and therefore superior to anything involving the risk of dirty fingernails. Also, as a woman in my early forties, I conformed to no right-minded picture of an apprentice farmer. And finally, with some chagrin I'll admit that I grew up among farmers and spent the first decades of my life plotting my escape from a place that seemed to offer me almost no potential for economic, intellectual, or spiritual satisfaction.

It took nigh onto half a lifetime before the valuables I'd casually left behind turned up in the lost and found.

The truth, though, is that I'd kept some of that treasure jingling in my pockets all along: I'd maintained an interest in gardening, always, dragging it with me wherever I went, even into a city backyard where a neighbor who worked the night shift insisted that her numerous nocturnal cats had every right to use my raised vegetable beds for their litter box. (I retaliated, in my way, by getting a rooster who indulged his right to use the hour of 6 A.M. for *his* personal compunctions.) In graduate school I studied ecology and evolutionary biology, but the complex mathematical models of predator-prey cycles only made sense to me when I converted them in my mind to farmstead analogies—even though, in those days, the Ecology Department and the College of Agriculture weren't on speaking terms. In my twenties, when I was trying hard to reinvent myself as a person without a Kentucky accent, I often found myself nevertheless the lone argumentative voice in social circles where "farmers" were lumped with political troglodytes and devotees of All-Star wrestling. Once in the early eighties, when cigarette smoking had newly and drastically fallen from fashion, I stood in someone's kitchen at a party and listened to something like a Greek Chorus chanting out the reasons why tobacco should be eliminated from the face of the earth, like smallpox. Some wild tug on my heart made me blurt out: "But what about the tobacco farmers?"

"Why," someone asked, glaring, "should I care about tobacco farmers?"

I was dumbstruck. I couldn't form the words to answer: yes, it is carcinogenic, and generally grown with too many inputs, but tobacco is the last big commodity in America that's still mostly grown on family farms, in an economy that won't let these farmers shift to another crop. If it goes extinct, so do they.

I couldn't speak because my mind was flooded with memory, pictures, scents, secret thrills. Childhood afternoons spent reading Louisa May Alcott in a barn loft suffused with the sweet scent of aged burley. The bright, warm days in late spring and early fall when school was functionally closed because whole extended families were drafted to the cooperative work of setting, cutting, stripping, or hanging tobacco. The incalculable fellowship measured out in funerals, family reunions, even bad storms or late-night calvings. The hard-muscled pride of showing I could finally throw a bale of hay onto the truckbed myself. (The year before, when I was eleven, I'd had the less honorable job of *driving* the truck.) The satisfaction of walking across the stage at high school graduation in a county where my name and my relationship to the land were both common knowledge.

But when pressed, that evening in the kitchen, I didn't try to defend the poor tobacco farmer. As if the deck were not already stacked against his little family enterprise, he was now tarred with the brush of evil along with the companies that bought his product, amplified its toxicity, and attempted to sell it to children. In most cases it's just the more ordinary difficulty of the small family enterprise failing to measure up to the requisite standards of profitability and efficiency. And in every case, the rational arguments I might frame in its favor will carry no weight without the attendant silk purse full of memories and sighs and songs of what family farming is worth. Those values are an old currency now, accepted as legal tender almost nowhere.

I found myself that day in the jaws of an impossible argument, and I find I am there still. In my professional life I've learned that as long as I write novels and nonfiction books about strictly human conventions and constructions, I'm taken seriously. But when my writing strays into that muddy territory where humans are forced to own up to our dependency on the land, I'm apt to be declared quaintly irrelevant by the small, acutely urban clique that decides in this country what will be called worthy literature. (That clique does not, fortunately, hold much sway over what people actually read.) I understand their purview, I think. I realize I'm beholden to people working in urban centers for many things I love: they publish books, invent theater, produce films and music. But if I had not been raised such a polite southern girl, I'd offer these critics a blunt proposition: I'll go a week without attending a movie or concert, you go a week without eating food, and at the end of it we'll sit down together and renegotiate "quaintly irrelevant."

This is a conversation that needs to happen. Increasingly I feel sure of it; I just don't know how to go about it when so many have completely forgotten the genuine terms of human survival. Many adults, I'm convinced, believe that food comes from grocery stores. In Wendell Berry's novel *Jayber Crow,* a farmer coming to the failing end of his long economic struggle despaired aloud, "I've wished sometimes that the sons of bitches would starve. And now I'm getting afraid they actually will."

Like that farmer, I am frustrated with the imposed acrimony between producers and consumers of food, as if this were a conflict in which one could possibly choose sides. I'm tired of the presumption of a nation divided between rural and urban populations whose interests are permanently at odds, whose votes will always be cast different ways, whose hearts and minds share no common ground. This is as wrong as blight, a useless way of thinking, similar to the propaganda warning us that any environmentalist program will necessarily be anti-human. Recently a national magazine asked me to write a commentary on the great divide between "the red and the blue"—imagery taken from election-night TV coverage that colored a map according to the party each state elected, suggesting a clear political difference between the rural heartland and urban coasts. Sorry, I replied to the magazine editors, but I'm the wrong person to ask: I live in red, tend to think blue, and mostly vote green. If you're looking for oversimplification, skip the likes of me.

Better yet, skip the whole idea. Recall that in every one of those red states, just a razor's edge under half the voters likely pulled the blue lever, and vice versa—not to mention the greater numbers *everywhere* who didn't even show up at the polls, so far did they feel from affectionate towards any of the available options. Recall that farmers and hunters, historically, are more active environmentalists than many progressive, city-dwelling vegetarians. (And conversely, that some of the strongest land-conservation movements on the planet were born in the midst of cities.) Recall that we all have the same requirements for oxygen and drinking water, and that we all like them clean but relentlessly pollute them. Recall that whatever lofty things you might accomplish today, you will do them only because you first ate something that grew out of dirt.

We don't much care to think of ourselves that way—as creatures whose cleanest aspirations depend ultimately on the health of our dirt. But our survival as a species depends on our coming to grips with that, along with some other corollary notions, and when I entered a comfortable midlife I began to see that my kids would get to do the same someday, or not, depending on how well our species could start owning up to its habitat and its food chain. As we faced one environmental crisis after another, did our species seem to be making this connection? As we say back home, Not so's you'd notice.

If a middle-aged woman studying agriculture seems strange, try this on for bizarre: most of our populace and all our leaders are participating in a mass hallucinatory fantasy in which the megatons of waste we dump in our rivers and bays are not poisoning the water, the hydrocarbons we pump into the air are not changing the climate, overfishing is not depleting the oceans, fossil fuels will never run out, wars that kill masses of civilians are an appropriate way to keep our hands on what's left, we are not desperately overdrawn at the environmental bank, and *really*, the kids are all right.

Okay, if nobody else wanted to talk about this, I could think about it myself and try to pay for my part of the damage, or at least start to tally up the bill. This requires a good deal of humility and a ruthless eye toward an average household's confusion between *need* and *want*. I reckoned I might get somewhere if I organized my life in a way that brought me face to face with what I am made of. The values I longed to give my children—honesty, cooperativeness, thrift, mental curiosity, physical competence—were intrinsic to my agrarian childhood, where the community organized itself around a sustained effort of meeting people's needs. These values, I knew, would not flow naturally from an aggressive consumer culture devoted to the sustained effort of inventing and engorging people's wants. And I could not, as any parent knows, prohibit one thing without offering others. So we would start with the simple and obvious: eschewing fast food for slow food, with the resulting time spent together in the garden and kitchen regarded as a plus, not a minus. We would skip TV in favor of interesting family work. We would participate as much as possible in the production of things our family consumes and the disposal of the things we no longer need. It's too easy to ignore damage you don't see and to undervalue things you haven't made yourself. Starting with food.

Meal preparations at our house, then, would not begin with *products,* like chicken tenders and frozen juice concentrate, but with whole things, like a chicken or an apple. A chicken or apple, what's more, with a background we could check up on. Our younger daughter was only a toddler when we first undertook this enterprise, but she seemed to grasp the idea. On a family trip once when we ate in a Chinese restaurant, she asked skeptically, "What was this duck's last name?"

What began as a kind of exercise soon turned into a kind of life, which we liked surprisingly well. It's enough to turn your stomach, anyway, to add up the fuel, money, and gunk that can go into food that isn't even *about* food. Our gustatory industries treat food items like spoiled little celebrities, zipping them around the globe in luxurious air-conditioned cabins, dressing them up in gaudy outfits, spritzing them with makeup, and breaking the bank on advertising, for heaven's sake. My farm-girl heritage makes me blush and turn down tickets to that particular circus. I'd rather wed my fortunes to the sturdy gal-next-door kind of food, growing what I need or getting it from local "you-pick" orchards and our farmers' market.

In making the effort to get acquainted with my food chain, I found country lanes and kind people and assets I had not known existed in my community. To my amazement, I found a CSA grower sequestered at the end of a dirt road within walking distance of my house, and he helped me fix an irrigation problem that had stumped me for months. I found others who would help me introduce a gardening program into my children's elementary school. I befriended the lone dairyman in my county who refuses to give hormones to his cows, not because he's paid more for the milk he sells to the cooperative (he isn't), but because he won't countenance treating his animals that way. I learned about heritage breeds, and that one of the rarest and tastiest of all turkeys, the Bourbon Red, was first bred a stone's throw from my hometown in Kentucky. I've come to know this bird inside and out, and intend to have my own breeding flock of them. I've become part of a loose-knit collective of poultrywomen who share tools and recipes and, at the end of the day, know how to make a real party out of harvest time. All in the house that good food built.

There is more to the story. It has come to pass that my husband and I, in what we hope is the middle of our lives, are in possession of a farm. It's not a hobby homestead, it is a *farm*, somewhat derelict but with good potential. It came to us with some twenty acres of good, tillable bottomland, plus timbered slopes and all the pasture we can ever use, if we're willing to claim it back from the brambles. A similar arrangement is available with the seventy-five-year-old apple orchard. The rest of the inventory includes a hundred-year-old clapboard house, a fine old barn that smells of aged burley, a granary, poultry coops, a root cellar, and a century's store of family legends. No poisons have been applied to this land for years, and we vow none ever will be.

I've never loved any earthly thing so much. It seems to my husband and me that this farm is something we need to work hard to deserve. As a former tobacco farm, it had a past without a future. But now that its future is in our hands, we recognize that it ought to feed people—more than just our family and those who come to our table. Precisely because of tobacco's changing fortunes, we're now situated in a community of farmers who are moving with courage and good cheer into the production of a regionally distributed line of organic produce. This economic project may be small in the eyes of global capitalism but it concerns us greatly, for its success or failure will be felt large in our schools, churches, and neighborhood businesses, not to mention our soil and streams, as these farmers make choices and, I hope, remain among us on their land. My family hopes to contribute to the endeavor as best we can, as producers as well as consumers, though with regard to the former we acknowledge our novice status. For several years now we've received from each other as gifts, on nearly all occasions, such books as are written by Gene Logsdon, Michael Phillips, Elliot Coleman, Carol Ekarius, Vandana Shiva, Wendell Berry. Some other wife might wish for diamond earrings, but my sweetheart knew I wanted *Basic Butchering*.

Our agrarian education has come in as a slow undercurrent beneath our workaday lives and the rearing of our children. Only our closest friends, probably, have taken real notice of the changes in our household: that nearly all the food we put on our table, in every season, was grown in our garden or very nearby. That the animals we eat took no more from the land than they gave back to it, and led sunlit, contentedly grassy lives. Our children know how to bake bread, stretch mozzarella cheese, ride a horse, keep a flock of hens laying, help a neighbor, pack a healthy lunch, and politely decline the world's less wholesome offerings. They know the first fresh garden tomato tastes as good as it does, partly, because you've waited for it since last Thanksgiving, and that the awful ones you could have bought at the grocery in between would only subtract from this equation. This rule applies to many things beyond tomatoes. I have noticed that the very politicians who support purely market-driven economics, which favor immediate corporate gratification over long-term responsibility, also express loud concern about the morals of our nation's children and their poor capacity for self-restraint. I wonder what kind of tomatoes those men feed their kids.

I have heard people of this same political ilk declare that it is perhaps sad but surely inevitable that our farms are being cut up and sold to make nice-sized lawns for suburban folks to mow, because the most immediately profitable land use must prevail in a free country. And yet I have visited countries where people are perfectly free, such as the Netherlands, where this sort of disregard for farmland is both illegal and unthinkable. Plenty of people in this country, too, seem to share a respect for land that gives us food; why else did so many friends of my youth continue farming even while the economic prospects grew doubtful? And why is it that more of them each year are following sustainable practices that defer some immediate profits in favor of the long-term health of their fields, crops, animals, and watercourses? Who are the legions of Americans who now allocate more of their household budgets to food that is organically, sustainably, and locally grown, rather than buying the cheapest products they can find? My husband and I, bearing these trends in mind, did not contemplate the profitable option of subdividing our farm and changing its use. Frankly, that seemed wrong.

It's an interesting question, how to navigate this tangled path between money and morality: not a *new* question by any means, but one that has taken strange turns in modern times. In our nation's prevailing culture there exists right now a considerable confusion between prosperity and success—so much so that avarice is frequently confused with a work ethic. One's patriotism and good sense may be called into doubt if one elects to earn less money or own fewer possessions than is humanly possible. The notable exception is that a person may do so for religious reasons: Christians are asked by conscience to tithe or assist the poor; Muslims do not collect interest; Catholics may respectably choose a monastic life of communal poverty; and any of us may opt out of a scheme that we feel to be discomforting to our faith. It is in this spirit that we, like you perhaps and so many others before us, have worked to rein in the free market's tyranny over our family's tiny portion of America and install values that override the profit motive. Upon doing so, we receive a greater confidence in our children's future safety and happiness. I believe we are also happier souls in the present, for what that is worth. In the darkest months I look for solace in seed catalogs and articles on pasture rotation. I sleep better at night, feeling safely connected to the things that help make a person whole. It is fair to say this has been, in some sense, a spiritual conversion.

Before I had read this book you're now holding in your hands, I would have hesitated to suggest that one's relationship to the land, to consumption and food, is a religious matter. But it's true; the decision to attend to the health of one's habitat and food chain is a spiritual choice. It's also a political choice, a scientific one, a personal and a convivial one. It's not a choice between living in the country or the town; it is about understanding that every one of us, at the level of our cells and respiration, lives in the country and is thus obliged to be mindful of the distance between ourselves and our sustenance.

I have worlds to learn about being a good farmer. Last spring when a hard frost fell upon our orchards on May 21st, I felt despair of ever getting there at all. But in any weather I may hope to carry a good agrarian frame of mind into my orchards and fields, my kitchen, my children's schools, my writing life, my friendships, my grocery shopping, and the county landfill. That's the point: it goes everywhere. It may or may not be a movement—I'll leave that to others to say. But it does move, and it works for us.

Introduction

Why Agrarianism Matters—Even to Urbanites

Norman Wirzba

It would seem, given the massive and unprecedented migration of farmers to urban centers, that a book on agrarianism is out of step with the times. After all, once independent farms are being consolidated into a few corporate conglomerates run by efficiency-minded, bottom-line agribusiness professionals. Driving through the American heartland shows that farming communities have become ghost towns, and consulting the Census Bureau demonstrates that farmers themselves have become a statistically irrelevant group. Picturesque farmyards, with their red barns and free-ranging chickens and geese, though having considerable storybook and advertising/marketing value, are in fact little more than quaint relics of a bygone era. Indeed, many would argue that former farmers are much better off working in cities, freed of the supposed drudgery and mind-numbing work of the farm and presumably able to depend on a steady wage, regular vacation days, and a secure pension. Overall, the mass movement from country to city represents a net gain.

Or does it? There are good reasons to suggest that a culture loses its indispensable moorings, and thus potentially distorts its overall aims, when it foregoes the sympathy and knowledge that grows out of cultivating (*cultura*) the land (*ager*). Past cultures needed to be attentive to the requirements of regional geographies because for thousands of years human life and development were themselves firmly and practically oriented through multiple relationships with natural landscapes and the organisms they support. Most people, as a matter of practical necessity, understood themselves and their aspirations in terms of the limits and possibilities of the land. And so, whether we appreciate it or not, current widespread insularity from and ignorance about our many interdependencies with the earth represents an unparalleled development in human history. Our novel situation requires us to consider if we can enact a vibrant and authentic culture without at the same time advocating a healthy *agri-culture*.

The purveyors of the industrial, and now information and global, economies routinely claim the inevitability and necessity of their programs and plans, and then argue that agrarian ways are anachronistic, even dangerous, since they stand in the way of a bright future. History shows, however, that the prophets and salespeople of technological progress rarely reveal the whole story. They do not tell us what the complete and long-term costs (to communities and ecosystems) of their visions are, when and where their visions fail, nor will they disclose the actual or potential profits they hope to realize. They hide the gap that often exists between promise and fulfillment.

Consider the following: (1) New technologies, first the television but now computers and the Internet, have been widely heralded as the indispensable tools to a more educated and intelligent democracy. It is plain to see, however, that democratic participation, civic responsibility, and general intelligence (including clear, honest communication) have not risen. Indeed, the evidence would suggest that corporate influence, which controls much of our technological media, has taken the place of democratic participation. (2) Nuclear power was promoted as the energy source that would make electricity safe, reliable, and "too cheap to meter." We now live with multi-billion dollar costs that must go toward securing or repairing existing facilities and cleaning up hazardous waste, hoping that this waste does not get into the wrong hands or into our groundwater and food systems. (3) The information economy would at last free us from the grime and labor of the industrial revolution, all the while lessening our dependence on diminishing natural resources. While it is true that many people no longer work manual labor jobs, it is not at all clear that labor has become less intensive, or that the pressure on natural habitats has eased. Indeed, frenetic consumption patterns mean that we are working more, even as our perception of the quality of life (for ourselves and for natural habitats) has gone down. (4) More recently we have been promised that biotechnology engineers will design organisms that will provide the world with a safe, steady, and secure source of food and pharmaceuticals. However, we are already beginning to see that genetically altered species (sometimes referred to as biological pollutants) pose grave risks to the stability of the ecosystems in which they were meant to flourish and on which they depend. (5) Finally, we are told that multinational corporations and global, centralized networks of business, communications, transportation, and production will raise the standard of life worldwide. Without minimizing the benefits that have accrued to some, it would be naive to overlook the select few growing extremely wealthy because of globalization, the corporate interests taking the place of national sovereignty and regional self-determination, the indigenous cultures and ways disappearing at an alarming rate, and the march of (mostly) Western power and money creating international tension and unrest.

Clearly it would be foolish to eschew all technological developments as uniformly bad, just as it would be ungrateful to overlook the many good things that have resulted in the last several decades. But in our haste to embrace technological improvements we must be careful not to overlook or degrade those elements of life—such as communal support, traditional wisdom, clean water, and nutritious food—that are fundamental. The long-term costs of some of our desires, when we take the time to measure them, may simply be too high to bear. While we (perpetually) search for a better tomorrow and a more luxurious place (somewhere else), we overlook the status of our homes and neighborhoods today. Here we can see that the social, economic, cultural, and environmental indicators are not very promising. In fact, many of us are tired of hearing the litanies of cultural and environmental problems: communal disintegration, social boredom and anxiety, nickel-and-dime employment, voter disenchantment, a growing gap between rich and poor, international terrorism and unrest, corporate welfare, biological/genetic pollution (super pests and pathogens), species loss, global warming, resource depletion, soil erosion, water and air degradation . . . and so on. This, rather than some hypothetical future, is our reality.

We should ask if these problems follow from the cultural mainstream as a direct consequence of the manners and goals prescribed by it. If they do, then we must also realize that it is unlikely that an enduring solution will arise from within the mainstream's midst. In other words, the cultural paradigm that causes or abets the crisis is unlikely to find a solution, primarily because the prevailing paradigm cannot see itself and its manners as the source of the problem. For that we must turn to people working at the margins of our culture, to those operating with a set of assumptions different from the prevailing order, to see what they think makes for a vibrant culture and natural environment.

Agrarianism is this compelling and coherent alternative to the modern industrial/technological/economic paradigm. It is not a throwback to a never-realized pastoral arcadia, nor is it a caricatured, Luddite-inspired refusal to face the future. It is, rather, a deliberate and intentional way of living and thinking that takes seriously the failures and successes of the past as they have been realized in our engagement with the earth and with each other. Authentic agrarianism, which should not be confused with farming per se (since severe economic pressure and the dash for quick profits have often led farmers to compromise agrarian ideals), represents the sustained attempt to live faithfully and responsibly in a world of limits and possibilities. As such it takes seriously what we know (and still need to learn) about the earth—the scientific ecological principles that govern all life forms—and what we know about each other—the social scientific and humanistic disciplines that enrich human self-understanding.

Agrarianism tests success and failure not by projected income statements or by economic growth, but by the health and vitality of a region's entire human and nonhuman neighborhood. Agrarianism, we might say, represents the most complex and far-reaching accounting system ever known, for according to it success must include a vibrant watershed and soil base; species diversity; human and animal contentment; communal creativity, responsibility, and joy; usable waste; social solidarity and sympathy; attention and delight; and the respectful maintenance of all the sources of life. Given the complexity and magnitude of this task, it is clear that authentic agrarianism has only been attempted thus far. Its full realization still awaits us. One of the primary aims of this book is to help us imagine and implement a genuinely agrarian vision.

The foregoing should make clear that agrarianism is not simply the concern or prerogative of a few remaining farmers, but is rather a comprehensive worldview that holds together in a synoptic vision the health of land *and* culture. What makes agrarianism the ideal candidate for cultural renewal is that it, unlike some environmental approaches that sequester wilderness and portray the human presence as invariably destructive or evil, grows out of the sustained, practical, intimate engagement between the power and creativity of both nature and humans. In agrarian practices we see a deliberate way of life in which the integrity and wholesomeness of peoples and neighborhoods, and the natural sources they depend upon, are maintained and celebrated. Agrarianism builds on the acknowledgment that we are biological and social beings that depend on healthy habitats and communities. However much we might think of ourselves as post-agricultural beings or disembodied minds, the fact of the matter is that we are inextricably tied to the land through our bodies—we have to eat, drink, and breathe—and so our culture must always be sympathetic to the responsibilities of agriculture. If we despise the latter, we are surely only a step away from despising the former too.

Urban Agrarian Culture

For too long, dominant worldviews have presupposed that we can design cultures without concern for the land's integrity and have naively taken it for granted that soils, waterways, and forests are simply resources to feed cultural ambition. Civilizations have presumed that greatness could be built at the expense of ecological degradation. The result, as can be seen in the cases of Sumer, the Mediterranean basin, and Mesoamerica, has invariably been either cultural ruin or severe cultural strain. What we need is a worldview that integrates and sees as continuous the limits and possibilities of land and culture together. What we need, as the economist Herman Daly has been saying for years, is a human economy attentive and responsive to the macro-economy of the land, the ecological patterns and possibilities that constitute our biological, physical, and chemical world. We must learn to put aside the misguided and destructive ambition that claims to achieve human health and happiness, while the natural habitats we live from and through languish and suffer.

The cultural neglect of natural habitats has had its corollary in the animosity between the country and the city, each side claiming for itself moral purity or human excellence. Farming folk have routinely described their way of life as conducive to peace, balance, and simple virtue, and the ways of the city as promoting strife, ambition, and greed. City folk, on the other hand, have considered cities as the entry into sophistication, creativity, and enlightenment, and farms as places of ignorance, provincialism, and limitation. Clearly this sort of antagonism, which is often based on caricatures, needs to be faced and overcome if we are to develop a synthetic vision promoting the health of land and culture together. We need to understand that the relationships between country and city have been varied and complex and, cross-border bickering notwithstanding, necessary. Can we together envision a culture that incorporates what is best from what we know about urban and rural life, and in so doing take steps that will ensure a vibrant and just future? Should we be contemplating an "urban agrarianism"?

A phrase like "urban agrarianism" indicates that our concern cannot simply be about the preservation of farmland, but must include the care of all living spaces—residential neighborhoods, schools and playgrounds, parks, and landfills, as well as glaciers, forests, wetlands, and oceans—the protection of all the places that maintain life. Farming has been central to the agrarian vision because farming practices, most obviously in food production, are our most direct and practical access to the processes of life and death. In them we learn about the limiting conditions of life, discover life's fragility and impermanence, but also life's giftedness and grace. As we eat and drink and breathe we visibly demonstrate, even if we do not always honor, our attachments to and dependence on the land. If we take care of the land and preserve the integrity of the soil base and watershed, we will at the same time insure the life contexts that are indispensable for cultural flourishing. If nothing else, we will at least demonstrate that we believe the future of our grandchildren is worth protecting.

Cultural practices, however, are not limited to eating and the production of food. We also strive to learn, to create, and to celebrate. If we are to do these well and responsibly, we must do them in ways that do not compromise the life-giving potential of communities and the land. And so while an agrarian vision must always have the integrity of energy/food flows as its foremost concern, its scope must also extend to the many places where we live and to the many tasks we perform. The reach of agrarian responsibilities is all-inclusive because all our activities, whether they occur in a steel and concrete office building, a commuter train, or a backyard garden, are informed and made possible by natural cycles of life and death. Every member and every moment on life's way are joined together in a bewildering ecological maze of cause and effect. The urbanite no less than the farmer is implicated in this web and so must appreciate the requirements and the costs for living things. To fail to do this is to risk ecological and cultural ruin.

The two forms of ruin go together. Plato recognized long ago that when a culture amasses luxury goods, or defines success primarily in terms of lavish consumption, the conditions for jealousy, enmity, and (violent) exploitation are set. What he did not fully appreciate is that the exploitation characteristic of a materialistic consumer culture is premised on the exploitation and exhaustion of nature's supporting habitats. All consumption, whether it is luxurious or not, takes place at the expense or sacrifice of habitats and other organisms. Given a finite resource base (its finitude becoming more apparent day by day as we note rates of deforestation, water scarcity, desertification, and suburban sprawl), it is simply prudent to limit our consumption to the scale of appropriate need rather than inappropriate wants, recognizing that propriety has as much to do with natural as it has to do with social limits.

Living in an urban context makes the adoption of agrarian responsibilities more difficult because as urbanites we do not, for the most part, feel or see our attachments to the land. We live increasingly in built environments that reflect human desires and ingenuity rather than natural limits and possibilities. We live at a blistering pace that is oblivious to the rhythms of natural cycles. And perhaps most important, we tend to act in complete ignorance of the effects our choices have on natural habitats. We do not appreciate or understand how actions as simple as placing garbage at the curb, purchasing a new computer, or adjusting the household thermostat can compromise or promote the natural contexts upon which we all depend. We tend not to connect increased electrical consumption with the leveling of whole mountains for their coal, or the discarding of used electronics with dangerously high accumulations of heavy metals and toxins in our groundwater. The effects are out of sight and thus out of mind. To think, however, that our actions are without effect is surely one of the great deceptions of our culture.

Agrarianism is about learning to take up the responsibilities that protect, preserve, and celebrate life. The first requirement of such responsibility is that we give up the delusion that we live in a purely human world of our own making, give up the arrogant and naive belief that human ambition should be the sole measure of cultural success or failure. As embodied beings we necessarily and beneficially draw our life from the many living beings that surround us and the natural processes that maintain them. If we are to care for these habitats we must, through education and direct experience, become attentive to our place in the wider natural world.

Agrarianism argues that attention and responsible action can occur most readily as we directly/practically see and feel our connections with each other and the land. For this reason, agrarians stress the importance of living as much as we can within local economies, economies that keep the loop between production and consumption as small as possible. Close communal contact and sustained commitment to a local, natural context increase the likelihood that our sight, feeling, and action are honest, nonevasive, and informed. If we can see how our living practices directly affect air and water quality, soil retention and health, species contentment and diversity, communal cohesion, and other markers of environmental health, and then learn to appreciate how nature's services enrich our personal and social lives, we will be much more inclined to change our practices in ways that benefit rather than bring harm to others. The assumption is that we are less likely to misuse or abuse the memberships we see benefiting us directly.

It is dangerous to romanticize local community life, especially when we remember that local communities have often been susceptible to various forms of provincialism. Farming communities, for instance, have not always been respectful of the contributions of women. Nor have they been very welcoming of foreigners or people with new ideas. The result has often been a form of communal claustrophobia. For good reason, then, when the opportunity for urban freedom, adventure, and anonymity arose, many young farmers took it. Wage employment, besides granting release from the steady demands of farm life, provided a relatively secure financial future, and it confined work to discrete hours of the day and week.

We now know, if we did not know before, that such unencumbered freedom carries with it the potential for considerable irresponsibility and destructiveness. We cannot live well—as friends, spouses, or citizens—if we do not respect and strengthen the bonds of relationship (human and nonhuman) that make life meaningful. Moreover, we should acknowledge that contemporary trends like globalization, corporate downsizing/restructuring, and movable capital have made all of us much less secure in our economic being. One of the hallmarks of postmodern life is its precariousness: we can no longer take our socioeconomic position for granted; we feel uncertain about the stability of the things and relationships we care most about; and we feel generally unsafe in the face of terrorism, vandalism, rape, and theft. It is no accident that literature on the themes of "home" and "community" is growing by leaps and bounds, as more and more people are seeking to ground their existence in something that is durable, safe, and life promoting.

Given these circumstances, it may well be that a broad-based discussion on the nature of responsible freedom should be our highest cultural priority. Can we envision and implement lives that encourage creativity, exploration, and self-expression and at the same time promote both the health of the natural habitats we live from and the vibrancy of communal structures that infuse personal life with meaning and joy? How will we maintain and celebrate the bonds of relationship that nurture life, without coming to regard these bonds as oppressive? These are complex, difficult questions that do not yield a simple solution applicable to all places at all times. Answering them, however, will require that we begin with a comprehensive and honest look at *where we are* as well as *who we are*, a thorough accounting of the costs of our desires and actions. It will also demand the creation of venues for sustained, engaged public discussion of these issues, especially when we recognize that democratic participation in public matters has been declining steadily .

Perhaps unsurprisingly, it is the very precarious nature of global/postmodern life that compels us to take our commitments lightly and to value our relationships less than we should. Movable capital demands a movable, and above all flexible, workforce. To get ahead in this world we must be ready to forsake any and all strategies/commitments to meet the new opportunities awaiting us. In this fluid context many of the bonds that tie us to community and place come to be treated like any consumable good—they can always be discarded if a new and potentially better "bond" comes along.

Clearly this trend must be resisted if we are to become attentive and affectionate caregivers of the places and communities we call home. It will not be overcome, however, as long as we remain wedded to the ambitions of our prevailing paradigm, ambitions that we know to be conducive to ill health, anxiety, stress, and fatigue. We need to see that the purveyors of this paradigm do not have our well-being at heart, but instead have everything to gain (financially) by keeping us unhappy, dissatisfied, and disengaged consumers. As we resist the ways of corporate and global ambition, we may yet come to see the grace and joy that accompany genuine efforts to make of our living places an enduring and convivial home. We will discover that our lives are everywhere maintained and benefited by the countless contributions of traditions, communities, habitats, and other organisms, and that we in turn have the potential to similarly benefit others.

To be an agrarian is to believe that we do not need the hypothetical (often false, and perpetually deferred) promises of a bright economic future to be happy and well. What we need—fertile land, drinkable water, solar energy, communal support and wisdom—we already have, or could have, if we turned our attention and energy to the protection and celebration of the sources of life.

We Are What/How We Eat

We can better understand the urgency and relevance of agrarian concerns and priorities if we consider the example of food security. It would seem, especially given the abundance and relative cheapness of food, that we do not have a food problem. The appearance, however, is deceiving. A growing number of farmers, ecologists, economists, and policy analysts are beginning to see that the complete costs associated with current food abundance are extremely high and that current pricing hides these costs from consumers. Food, for the most part, is now an industrial product. As such its character and quality, as well as the conditions under which it is produced, are determined by the demands of industrial and market efficiency. While this might make good economic sense, the effect of treating food as an industrial rather than as a natural and cultural product has been the abuse of land, animals, and human communities.

Consider first the fact that farmer independence is a thing of the past. The food system, which includes food being grown, harvested, processed, packaged, distributed, shipped, and marketed, has increasingly come under the control and ownership of a small number of giant corporations like Cargill, Archer-Daniels-Midland Co., Monsanto, and ConAgra. These companies subscribe to "vertical integration," which means they play a determining role in all the processes of food production and distribution, providing the consumer with products they control from "farm gate to dinner plate." They determine what is planted or fed, how much, by whom, and at what price. In this scenario, farmers are reduced to serfs. They must sign a contract with the company at the beginning of the season and stick to the demands of the corporation. In many cases farmers assume all the risk and liability. They must also go to the same company (or one of its partners) for the many expensive farm inputs—seed, fertilizer, insecticides/pesticides—that are essential for profitability. Farmers who refuse to sign these contracts often find that simply no market exists for their product, since the same company or its subsidiary controls purchasing and distribution. The goal is to make farmers completely beholden to corporate interests and profitability. The most extreme example of corporate control can be seen in the invention of the "terminator gene," a genetically modified plant that produces sterile seed. The effect of this biotechnological advance is that farmers cannot save their own seed for next year's planting (thereby saving considerable sums of money, as well as securing farm self-sufficiency), but must purchase their seed at the store each year.

Is this transformation of the food system a good idea for farmers, communities, consumers, or the land? Clearly it has not been good for farmers. While corporate profits in the food sector have soared over the last several years, small farmers have seen a steady decline in income. Desperate for any margin of profitability, many farmers embrace whatever mechanical, biotechnological, agrochemical product they can, with the net effect that corporate profitability again increases at the farmers' expense. The farmers who survive must either get really big or get out, hoping they will have enough cash flow to get them through another bad year in which the costs of production exceed product income. For good reason, observers of the contemporary agricultural scene suggest that rural communities look more and more like mining communities. Everything of value is sucked out by the corporate office.

Rural communities have suffered greatly as a result of this transformed food system. With the demise of local seed companies, local purchasers, and processors and distributors, money that would have circulated several times within a community (and thus benefited many businesses and families) goes elsewhere. With this cash exodus, small towns and cities that were once the heart of American cultural life find it impossible to maintain basic services in education, health care, construction, and general social welfare. There is no place to go but the big city. On the other hand, megafarm managers are increasingly hiring minimum-wage, migrant workers, resulting in rural ghettoes replete with social problems similar to their urban counterparts.

Rural communities also bear the brunt of noxious corporate farming practices. While taxpayers absorb the costs of tax incentives and price subsidies to induce big producers to set up shop in their states or counties, local communities must deal with disgusting odors, contaminated ground and surface water, accumulated toxic waste, and stressed infrastructure mechanisms like roadways and waterways. These costs are rarely picked up by the producers responsible for them.

As consumers we should be asking whether or not the free exchange of products, the stewardship of public goods like soil and water, or more fundamental yet, informed public discussion about food issues can result from a context where integrated corporate monopolies set pricing and production. Consumers are mostly ignorant about how food is produced and provided, so they are in no position to understand, let alone confront, agricultural abuses like the depletion or contamination of public water supplies or the heavy use of antibiotics and hormones in meat and dairy operations. Doctors are increasingly aware that public health costs will increase dramatically as we confront super pests and viruses that evolve in confined farm factories. The costs of cleaning up water contaminated by agricultural runoff will also need to be picked up by consumers. Moreover, industrial farming stresses monocultures, which means the growing of one crop variety on a vast acreage (unbroken fields of wheat, corn, or rice), or the breeding of livestock from a very limited gene pool (90 percent of all commercially produced turkeys, for instance, come from just three breeding flocks). Monocultures of this kind are highly susceptible to disease and pest infestation. The threat of species collapse can be held at bay only with ever more toxic and expensive pesticides.

We cannot solve this problem of contaminated or compromised habitats simply by moving it to another region or country. Recently some economists have suggested that we should shift food production almost entirely to developing nations where food can be grown more cheaply (and where the hazards and costs of industrial production can be absorbed by others). Farmland in developed countries can thus be freed to the bucolic pleasures of the wealthy. This view, besides being unjust, is ecologically naive. It is foolish not to have a diversified food production network, a network spread out over many regions, if for no other reasons than decreased likelihood of massive crop failure and increased ability to capitalize on the strengths certain regions possess for the growth of particular crops. The old adage that one should never put all of one's eggs in one basket is especially true here. A stable food system, much like a stable and resilient habitat, depends on a diversity of crops grown over diverse landscapes. Food diversity attuned to regional ecological possibilities, rather than the massive monocultures of today, is our best defense against foreign attack, whether it comes from pests or terrorists. History has shown repeatedly that as regions grow and consume their own food and rely as little as possible on food imports, their food supply becomes more secure.

The most dangerous and pervasive threat to our food system, however, may well be the exhaustion or destruction of our land base. Agricultural experts are now discovering that returns on chemically intensive farming are actually decreasing. Many factors contribute to this decrease, foremost among them the degradation of the soil base itself. Besides massive erosion rates (in some regions one bushel of crop is matched by two to five bushels of soil lost to erosion), the heavy application of synthetic fertilizers and pesticides has reduced fertile soils teeming with billions of microorganisms to zombie status. We kill these life-promoting microorganisms and deplete the soil's organic content, thereby reducing it to lifeless dirt incapable of fending off pests or supporting plant life without synthetically produced (and fossil fuel derived) chemical additives. We are losing quality topsoil at a rate far faster than it can be replenished.

Industrial farming is heavily dependent on cheap fossil fuel, not only for farm power, but also for fertilizer/pesticide production, irrigation, and food transport (the average grocery store item travels thirteen hundred to fifteen hundred miles before it reaches the shelf) and preparation. In some instances it takes ten calories of fossil fuel energy to produce one calorie of food energy. Is this not grossly inefficient and destructive of our habitats and health? Clearly, this agricultural malpractice can be sustained only because of a greater malfunctioning in our energy sector that greatly subsidizes the fossil fuel industry, going so far as to protect it by means of war. The tragedy of this scenario is that farmers and consumers would not need to pay many of the exorbitant costs associated with industrial production if we could learn to work more attentively with natural processes that can provide many food services like pest control, water retention, soil fertility, and solar heating and energy for free.

Given this partial list of problems, we now need to ask if a food system can be secure if it depends on making its farmers, communities, consumers, and land base insecure. Our highly centralized food system, besides being antidemocratic, hugely wasteful, and destructive, is also vulnerable to external threats of terrorism, volatile global markets, and pests (because monocultures have the weakest immunity in nature). How would an agrarian worldview address this situation of insecurity?

For starters, agrarians would not propose that we all become subsistence farmers growing all of our own food. But all people, because they eat and drink, should be aware of food responsibilities and take a more active role in promoting food safety and security. Food security is grounded in regional food production. Food responsibility begins with becoming educated about the food industry, learning about those products and processes that promote health and vitality for the entire food neighborhood (should we expect wholesome eating if plants and livestock are sickly or under stress?). Given recent revelations of unsavory corporate food production practices, having as much information about the food system as possible is a bare minimum. The best way to accomplish this education is to form relationships with food providers, as can be done when we purchase from farmers directly, at farmers' markets or through community supported agriculture (CSA) projects. The key is to become involved participants in food production wherever we can. Best yet, we can make the effort to grow some of our own food, and thus directly see what we are eating and under what conditions it was produced.

Food is the most direct link we have between culture and nature, city and farm folk. It can serve as the point of interest that unites urbanites with farming concerns. We need to see that the "plight of the farmer" does not affect simply farmers, but all of us. The preservation of soils and watersheds, humane treatment of animal livestock, worker safety and contentment, common ownership of the genetic food base (rather than its patented protection by biotech food firms)—these are objectives that we should all have in common. We need to support farmers who are dedicated to preserving healthy farmland, and to encourage our governments to stop their financial support of megafarms and the corporate interests they support. We need also to reward those growers who dare to be independent by giving them our commitment to purchase from them, even if that means paying higher prices for food. Above all we need to get past the idea that cheaper food is better food, especially when we remember that the cheapness of food is made possible by the externalization of many ecological and cultural (especially health) costs, costs that we will end up paying in some other way.

This is not an argument for expensive food. In fact, there are a number of ways that food can become much more affordable. The most obvious would be to grow some of our own. Urban agriculture, what some refer to as the "quiet revolution" in food production, is making it clear that food can be grown in a great variety of urban settings—in backyards, on rooftops, in window boxes and basements, on vacant lots, greenbelts, or playgrounds, and in public housing projects. Community garden plots can become gathering places for the growth of nutritious food and revitalized communities, as residents form relationships that support and help each other. Another way to reduce food costs would be to extricate food production from the expenses associated with industrial production methods that consume great quantities of fossil fuels and depend on extensive processing, marketing, and distribution networks. There is no reason why consumers and institutions (schools and hospitals, for instance) cannot buy more from local producers, and thus cut down on many costs. Their food will be fresher and more nutritious. Consumers, in turn, can benefit producers by giving to them their recyclable waste and their steady business. Local businesses that facilitate these transactions will become a communal priority, thus keeping more of our food dollars circulating in the communities where we live.

As we begin to understand that food is not simply fuel, but is in fact a natural, social, cultural, and spiritual product, we will also make the effort to foster the practical conditions necessary to protect and preserve ecological and social health. Our safety does not reside in the proxies we give to food corporations that premise their success on compromised habitats and communities, mistreated livestock, and market dominance. It rests rather in the responsible support and celebration of regional networks that join together producer and consumer, country and city, nature and culture.

READING *THE ESSENTIAL AGRARIAN READER*

The essays in this book have been grouped into three parts, each having a different emphasis. Part 1 describes and develops the key principles of an agrarian worldview. Wendell Berry's "The Agrarian Standard" demonstrates that the agrarian view is fundamentally at odds with the received industrial paradigm. Industrialism is the way of the machine, the way of technological invention that premises economic success on the exploitation of habitats and communities. Agrarianism, by contrast, is a way of life attuned to requirements of land and local communities. Historically speaking, those who have tried to be faithful to the land have had a very difficult time. Brian Donahue describes why this has been the case and then offers an agrarian framework that might guide and correct current land and community development. Building on Berry's assertion that we have been "unsettling America" from the start, Donahue offers creative suggestions for how we might begin to settle into our land at last.

Maurice Telleen takes on the issue of whether or not agrarianism, as an "ism," qualifies as a movement. After giving several reasons for why it might be best not to consider it that way, he shows how agrarianism aims to be a comprehensive value system that orders life in its various dimensions. Because such a value system can readily be compared to a "quasi-religion," Telleen then develops the Ten Agrarian Commandments, which might very well "save this country's bacon" (commandments of any sort have everything to do with the structuring of our economic lives). Herman Daly offers a careful assessment of why the prevailing economic paradigm is not genuinely sustainable. Indeed, given the inability of many economists to integrate in their calculations basic necessities like fertile soil, fresh/clean water, stable/resilient habitats, and ecological wisdom and sympathy, this essay offers a preliminary and indispensable challenge to economists to expand their range of consideration to include what Daly calls "ecological throughput." A full-blown agrarian economics, an economic system that is attuned to the wide sweep of natural and cultural concerns as we have so far described them, must build on the principles of sustainability as here described. Successful economic development depends on the acknowledgment of ecological inputs and limits and the promotion of policies that focus less on growth (which in some cases turns out to be uneconomic) and more on quality of life.

In the essay "Placing the Soul" I explain why it has been difficult for us as a culture to develop the habits that would promote responsible dwelling within the land. I argue that dominant philosophical and religious traditions, by stressing the preeminence of spiritual souls, have falsified our true nature as embodied beings that necessarily live through our connections with habitats and the lives of others. I conclude that a more authentic spiritual life is possible once we acknowledge and accept responsibility for the places and communities in terms of which we live.

Part 2 examines the current state of agriculture and its effects on broad cultural concerns, showing that problems in agriculture are reflective of cultural malfunction. The essay by Fred Kirschenmann describes the transformation of agriculture over the last several decades from relatively small, independent producers to the megafarms of today. He links this development to economic factors that have dramatically altered social and natural landscapes. He then considers public policy options that can help us avert the destructive ecological and social consequences following from the current paradigm. Vandana Shiva's essay shows that industrial farming is having a similarly destructive effect in India. As food corporations expand their global reach (Shiva uses the metaphor of war to express this development), local producers invariably suffer. Shiva shows that the promises of increased food productivity do not hold and that our best hope for global food security is to protect ecological and cultural diversity and equip local producers.

Wes Jackson describes how modern industrial farming rests on mistaken scientific assumptions. He argues that we need a natural systems agriculture, which patterns agricultural production on natural processes that have developed throughout evolutionary time. Authentic agrarian practices reflect an empathetic mind that is attentive to the limits and possibilities of geophysical places. Gene Logsdon's essay shows how this empathy might be practically realized through pasture farming. The predominant method of meat and dairy production is to confine huge numbers of livestock in a small area and then feed them corn, hormones, and antibiotics. Logsdon shows how this practice, besides being inhumane, is ecologically destructive and wasteful and can be replaced with range feeding that is much less expensive for farmers and healthier for animals.

The essay by David Orr looks more carefully at why an agrarian message has not been well received by the cultural mainstream. He shows how the assumptions that drive industrial agriculture also permeate society and thus prevent us from facing honestly the truth of who we are. We are, Orr says, a culture in denial of our material and biological contexts;

thus, we cannot possibly develop a healthy culture. Our best future depends on the development and implementation of an agrarian worldview that holds together the wholeness of habitats and communities.

The essays in part 3 illustrate some of the practical effects that would follow from an agrarian agenda. The essay by Benjamin Northrup and Benjamin Lipscomb shows that agrarian concerns and priorities bear a remarkable affinity with the concerns and priorities of the "New Urbanism," a movement in architecture and urban planning that promotes vibrant and viable neighborhoods. They show how agrarians and New Urbanists can learn from each other in the effort to make our living places a genuine home. Susan Witt then documents several particular strategies—notably the Community Land Trust and the implementation of local currencies—that have been tried and found to be successful in the preservation of farmland and the promotion of community projects. She highlights the specific challenges facing local business and farming development and argues for solutions that grow out of communal problem solving.

If we take the agrarian worldview seriously, believe it to be necessary for ecological and cultural health, it is imperative that we learn to put words and ideas into action. Hank Graddy documents the work the Sierra Club (among other environmental groups) has done to stop, or at least mitigate, the destructive impacts of industrial agriculture, most notably concentrated animal feeding operations (CAFOs). He shows how local communities can work together to legally and legislatively protect their food and water supplies and promote the humane treatment of agricultural livestock. Eric Freyfogle shows how an agrarian framework can help us reformulate basic notions about private property and land ownership. He begins by noting that the entailments and responsibilities of land ownership have changed through time and that we can work together to form ownership rights that are more attentive to ecological realities and communal necessities.

The final essay, "Going to Work," asks us to think more carefully and more broadly about the nature of work. In an age when career specialization and advancement have taken the place of vocational responsibility, Wendell Berry prompts us to reconsider the practical conditions necessary for us to make of our work an art that serves the health and well-being of the neighborhoods in which we live. More specifically, Berry draws the contrast between a sympathetic/affectionate mind and a mind that aspires only to (increasingly economic) reason, arguing that it will be through the former (though not entirely without the latter) mind that we may come to enact the virtues of humility, reverence, proper scale, and good workmanship.

Several of the essays in this volume made their first appearance (sometimes in rather different form) as speeches delivered at "The Future of Agrarianism: *The Unsettling of America* Twenty-Five Years Later," a 2002 conference hosted by Georgetown College. In addition, a few essays were written by authors who could not attend the conference but who, nonetheless, wanted to pay tribute to Wendell Berry, the author of *The Unsettling of America*. Besides marking the twenty-fifth anniversary of the publication of his book, the conference sought to extend the agrarian argument first set down there, to demonstrate that the ideas originally expressed had not outgrown their usefulness, but are now more compelling and necessary than ever.

It would be difficult to overestimate the significance of Wendell Berry's work for articulating and defending the agrarian cause. Through his writing, instruction, and personal example, Berry has been an inspiration and indispensable guide for many of us. This collection of essays is a vibrant testimony to the continuing relevance and usefulness of agrarian insights and practices. It is offered, in part, as an expression of gratitude for Wendell Berry's important contribution and as a stimulant for cultural reform. Our present cultural course is not inevitable. The great task before us is to envision and implement a better future. These essays present an agrarian alternative that may yet lead us, as Berry says, into "the grace of the world," the place where we will find our true freedom and joy.

Part 1

Agrarian Principles and Priorities

1

The Agrarian Standard

Wendell Berry

The Unsettling of America was published twenty-five years ago; it is still in print and is still being read. As its author, I am tempted to be glad of this, and yet, if I believe what I said in that book, and I still do, then I should be anything but glad. The book would have had a far happier fate if it could have been disproved or made obsolete years ago.

It remains true because the conditions it describes and opposes, the abuses of farmland and farming people, have persisted and become worse over the last twenty-five years. In 2002 we have less than half the number of farmers in the United States that we had in 1977. Our farm communities are far worse off now than they were then. Our soil erosion rates continue to be unsustainably high. We continue to pollute our soils and streams with agricultural poisons. We continue to lose farmland to urban development of the most wasteful sort. The large agribusiness corporations that were mainly national in 1977 are now global, and are replacing the world's agricultural diversity, which was useful primarily to farmers and local consumers, with bioengineered and patented monocultures that are merely profitable to corporations. The purpose of this new global economy, as Vandana Shiva has rightly said, is to replace "food democracy" with a worldwide "food dictatorship."[1]

To be an agrarian writer in such a time is an odd experience. One keeps writing essays and speeches that one would prefer not to write, that one wishes would prove unnecessary, that one hopes nobody will have any need for in twenty-five years. My life as an agrarian writer has certainly involved me in such confusions, but I have never doubted for a minute the importance of the hope I have tried to serve: the hope that we might become a healthy people in a healthy land.

We agrarians are involved in a hard, long, momentous contest, in which we are so far, and by a considerable margin, the losers. What we have undertaken to defend is the complex accomplishment of knowledge, cultural memory, skill, self-mastery, good sense, and fundamental decency—the high and indispensable art—for which we probably can find no better name than "good farming." I mean farming as defined by agrarianism as opposed to farming as defined by industrialism: farming as the proper use and care of an immeasurable gift.

I believe that this contest between industrialism and agrarianism now defines the most fundamental human difference, for it divides not just two nearly opposite concepts of agriculture and land use, but also two nearly opposite ways of understanding ourselves, our fellow creatures, and our world.

The way of industrialism is the way of the machine. To the industrial mind, a machine is not merely an instrument for doing work or amusing ourselves or making war; it is an explanation of the world and of life. The machine's entirely comprehensible articulation of parts defines the acceptable meanings of our experience, and it prescribes the kinds of meanings the industrial scientists and scholars expect to discover. These meanings have to do with nomenclature, classification, and rather short lineages of causation. Because industrialism cannot understand living things except as machines, and can grant them no value that is not utilitarian, it conceives of farming and forestry as forms of mining; it cannot use the land without abusing it.

Industrialism prescribes an economy that is placeless and displacing. It does not distinguish one place from another. It applies its methods and technologies indiscriminately in the American East and the American West, in the United States and in India. It thus continues the economy of colonialism. The shift of colonial power from European monarchy to global corporation is perhaps the dominant theme of modern history. All along—from the European colonization of Africa, Asia, and the New World, to the domestic colonialism of American industries, to the colonization of the entire rural world by global corporations—it has been the same story of the gathering of an exploitive economic power into the hands of a few people who are alien to the places and the people they exploit. Such an economy is bound to destroy locally adapted agrarian economies everywhere it goes, simply because it is too ignorant not to do so. And it has succeeded precisely to the extent that it has been able to inculcate the same ignorance in workers and consumers. A part of the function of industrial education is to preserve and protect this ignorance.

To the corporate and political and academic servants of global industrialism, the small family farm and the small farming community are not known, not imaginable, and therefore unthinkable, except as damaging stereotypes. The people of "the cutting edge" in science, business, education, and politics have no patience with the local love, local loyalty, and local knowledge that make people truly native to their places and therefore good caretakers of their places. This is why one of the primary principles of industrialism has always been to get the worker away from home. From the beginning it has been destructive of home employment and home economies. The office or the factory or the institution is the place for work. The economic function of the household has been increasingly the consumption of purchased goods. Under industrialism, the farm too has become increasingly consumptive, and farms fail as the costs of consumption overpower the income from production.

The idea of people working at home, as family members, as neighbors, as natives and citizens of their places, is as repugnant to the industrial mind as the idea of self-employment. The industrial mind is an organizational mind, and I think this mind is deeply disturbed and threatened by the existence of people who have no boss. This may be why people with such minds, as they approach the top of the political hierarchy, so readily sell themselves to "special interests." They cannot bear to be unbossed. They cannot stand the lonely work of making up their own minds.

The industrial contempt for anything small, rural, or natural translates into contempt for uncentralized economic systems, any sort of local self-sufficiency in food or other necessities. The industrial "solution" for such systems is to increase the scale of work and trade. It is to bring Big Ideas, Big Money, and Big Technology into small rural communities, economies, and ecosystems—the brought-in industry and the experts being invariably alien to and contemptuous of the places to which they are brought in. There is never any question of propriety, of adapting the thought or the purpose or the technology to the place.

The result is that problems correctable on a small scale are replaced by large-scale problems for which there are no large-scale corrections. Meanwhile, the large-scale enterprise has reduced or destroyed the possibility of small-scale corrections. This exactly describes our present agriculture. Forcing all agricultural localities to conform to economic conditions imposed from afar by a few large corporations has caused problems of the largest possible scale, such as soil loss, genetic impoverishment, and groundwater pollution, which are correctable only by an agriculture of locally adapted, solar-powered, diversified small farms—a correction that, after a half century of industrial agriculture, will be difficult to achieve.

The industrial economy thus is inherently violent. It impoverishes one place in order to be extravagant in another, true to its colonialist ambition. A part of the "externalized" cost of this is war after war.

Industrialism begins with technological invention. But agrarianism begins with givens: land, plants, animals, weather, hunger, and the birthright knowledge of agriculture. Industrialists are always ready to ignore, sell, or destroy the past in order to gain the entirely unprecedented wealth, comfort, and happiness supposedly to be found in the future. Agrarian farmers know that their very identity depends on their willingness to receive gratefully, use responsibly, and hand down intact an inheritance, both natural and cultural, from the past. Agrarians understand themselves as the users and caretakers of some things they did not make, and of some things that they cannot make.

I said a while ago that to agrarianism farming is the proper use and care of an immeasurable gift. The shortest way to understand this, I suppose, is the religious way. Among the commonplaces of the Bible, for example, are the admonitions that the world was made and approved by God, that it belongs to Him, and that its good things come to us from Him as gifts. Beyond those ideas is the idea that the whole Creation exists only by participating in the life of God, sharing in His being, breathing His breath. "The world," Gerard Manley Hopkins said, "is charged with the grandeur of God."[2] Such thoughts seem strange to us now, and what has estranged us from them is our economy. The industrial economy could not have been derived from such thoughts any more than it could have been derived from the Golden Rule.

If we believed that the existence of the world is rooted in mystery and in sanctity, then we would have a different economy. It would still be an economy of use, necessarily, but it would be an economy also of return. The economy would have to accommodate the need to be worthy of the gifts we receive and use, and this would involve a return of propitiation, praise, gratitude, responsibility, good use, good care, and a proper regard for the unborn. What is most conspicuously absent from the industrial economy and industrial culture is this idea of return. Industrial humans relate themselves to the world and its creatures by fairly direct acts of violence. Mostly we take without asking, use without respect or gratitude, and give nothing in return. Our economy's most voluminous product is waste—valuable materials irrecoverably misplaced, or randomly discharged as poisons.

To perceive the world and our life in it as gifts originating in sanctity is to see our human economy as a continuing moral crisis. Our life of need and work forces us inescapably to use in time things belonging to eternity, and to assign finite values to things already recognized as infinitely valuable. This is a fearful predicament. It calls for prudence, humility, good work, propriety of scale. It calls for the complex responsibilities of caretaking and giving-back that we mean by "stewardship." To all of this the idea of the immeasurable value of the resource is central.

We can get to the same idea by a way a little more economic and practical, and this is by following through our literature the ancient theme of the small farmer or husbandman who leads an abundant life on a scrap of land often described as cast-off or poor. This figure makes his first literary appearance, so far as I know, in Virgil's Fourth Georgic:

> I saw a man,
> An old Cilician, who occupied
> An acre or two of land that no one wanted,
> A patch not worth the ploughing, unrewarding
> For flocks, unfit for vineyards; he however
> By planting here and there among the scrub
> Cabbages or white lilies and verbena
> And flimsy poppies, fancied himself a king
> In wealth, and coming home late in the evening
> Loaded his board with unbought delicacies.[3]

Virgil's old squatter, I am sure, is a literary outcropping of an agrarian theme that has been carried from earliest times until now mostly in family or folk tradition, not in writing, though other such people can be found in books. Wherever found, they don't vary by much from Virgil's prototype. They don't have or require a lot of land, and the land they have is often marginal. They practice subsistence agriculture, which has been much derided by agricultural economists and other learned people of the industrial age, and they always associate frugality with abundance.

In my various travels, I have seen a number of small homesteads like that of Virgil's old farmer, situated on "land that no one wanted" and yet abundantly productive of food, pleasure, and other goods. And especially in my younger days, I was used to hearing farmers of a certain kind say, "They may run me out, but they won't starve me out" or "I may get shot, but I'm not going to starve." Even now, if they cared, I think agricultural economists could find small farmers who have prospered, not by "getting big," but by practicing the ancient rules of thrift and subsistence, by accepting the limits of their small farms, and by knowing well the value of having a little land.

How do we come at the value of a little land? We do so, following this strand of agrarian thought, by reference to the value of *no* land. Agrarians value land because somewhere back in the history of their consciousness is the memory of being landless. This memory is implicit, in Virgil's poem, in the old farmer's happy acceptance of "an acre or two of land that no one wanted." If you have no land you have nothing: no food, no shelter, no warmth, no freedom, no life. If we remember this, we know that all economies begin to lie as soon as they assign a fixed value to land. People who have been landless know that the land is invaluable; it is worth everything. Pre-agricultural humans, of course, knew this too. And so, evidently, do the animals. It is a fearful thing to be without a "territory." Whatever the market may say, the worth of the land is what it always was: It is worth what food, clothing, shelter, and freedom are worth; it is worth what life is worth. This perception moved the settlers from the Old World into the New. Most of our American ancestors came here because they knew what it was to be landless; to be landless was to be threatened by want and also by enslavement. Coming here, they bore the ancestral memory of serfdom. Under feudalism, the few who owned the land owned also, by an inescapable political logic, the people who worked the land.

Thomas Jefferson, who knew all these things, obviously was thinking of them when he wrote in 1785 that "it is not too soon to provide by every possible means that as few as possible shall be without a little portion of land. The small landholders are the most precious part of a state. . . ."[4] He was saying, two years before the adoption of our constitution, that a democratic state and democratic liberties depend upon democratic ownership of the land. He was already anticipating and fearing the division of our people into settlers, the people who wanted "a little portion of land" as a home, and, virtually opposite to those, the consolidators and exploiters of the land and the land's wealth, who would not be restrained by what Jefferson called "the natural affection of the human mind."[5] He wrote as he did in 1785 because he feared exactly the political theory that we now have: the idea that government exists to guarantee the right of the most wealthy to own or control the land without limit.

In any consideration of agrarianism, this issue of limitation is critical. Agrarian farmers see, accept, and live within their limits. They understand and agree to the proposition that there is "this much and no more." Everything that happens on an agrarian farm is determined or conditioned by the understanding that there is only so much land, so much water in the cistern, so much hay in the barn, so much corn in the crib, so much firewood in the shed, so much food in the cellar or freezer, so much strength in the back and arms—and no more. This is the understanding that induces thrift, family coherence, neighborliness, local economies. Within accepted limits, these virtues become necessities. The agrarian sense of abundance comes from the experienced possibility of frugality and renewal within limits.

This is exactly opposite to the industrial idea that abundance comes from the violation of limits by personal mobility, extractive machinery, long-distance transport, and scientific or technological breakthroughs. If we use up the good possibilities in this place, we will import goods from some other place, or we will go to some other place. If nature releases her wealth too slowly, we will take it by force. If we make the world too toxic for honeybees, some compound brain, Monsanto perhaps, will invent tiny robots that will fly about, pollinating flowers and making honey.

To be landless in an industrial society obviously is not at all times to be jobless and homeless. But the ability of the industrial economy to provide jobs and homes depends on prosperity, and on a very shaky kind of prosperity too. It depends on "growth" of the wrong things, such as roads and dumps and poisons—on what Edward Abbey called "the ideology of the cancer cell"—and on greed with purchasing power. In the absence of growth, greed, and affluence, the dependents of an industrial economy too easily suffer the consequences of having no land: joblessness, homelessness, and want. This is not a theory. We have seen it happen.

I don't think that being landed necessarily means owning land. It does mean being connected to a home landscape from which one may live by the interactions of a local economy and without the routine intervention of governments, corporations, or charities.

In our time it is useless and probably wrong to suppose that a great many urban people ought to go out into the countryside and become homesteaders or farmers. But it is not useless or wrong to suppose that urban people have agricultural responsibilities that they should try to meet. And in fact this is happening. The agrarian population among us is growing, and by no means is it made up merely of some farmers and some country people. It includes urban gardeners, urban consumers who are buying food from local farmers, organizers of local food economies, consumers who have grown doubtful of the healthfulness, the trustworthiness, and the dependability of the corporate food system—people, in other words, who understand what it means to be landless.

Apologists for industrial agriculture rely on two arguments. In one of them, they say that the industrialization of agriculture, and its dominance by corporations, has been "inevitable." It has come about and it continues by the agency of economic and technological determinism. There has been simply nothing that anybody could do about it.

The other argument is that industrial agriculture has come about by choice, inspired by compassion and generosity. Seeing the shadow of mass starvation looming over the world, the food conglomerates, the machinery companies, the chemical companies, the seed companies, and the other suppliers of "purchased inputs," have done all that they have done in order to solve "the problem of hunger" and to "feed the world."

We need to notice, first, that these two arguments, often used and perhaps believed by the same people, exactly contradict each other. Second, though supposedly it has been imposed upon the world by economic and technological forces beyond human control, industrial agriculture has been pretty consistently devastating to nature, to farmers, and to rural communities, at the same time that it has been highly profitable to the agribusiness corporations, which have submitted not quite reluctantly to its "inevitability." And, third, tearful over human suffering as they always have been, the agribusiness corporations have maintained a religious faith in the profitability of their charity. They have instructed the world that it is better for people to buy food from the corporate global economy than to raise it for themselves. What is the proper solution to hunger? Not food from the local landscape, but industrial development. After decades of such innovative thought, hunger is still a worldwide calamity.

The primary question for the corporations, and so necessarily for us, is not how the world will be fed, but who will control the land, and therefore the wealth, of the world. If the world's people accept the industrial premises that favor bigness, centralization, and (for a few people) high profitability, then the corporations will control all of the world's land and all of its wealth. If, on the contrary, the world's people might again see the advantages of local economies, in which people live, so far as they are able to do so, from their home landscapes, and work patiently toward that end, eliminating waste and the cruelties of landlessness and homelessness, then I think they might reasonably hope to solve "the problem of hunger," and several other problems as well.

But do the people of the world, allured by TV, supermarkets, and big cars, or by dreams thereof, *want* to live from their home landscapes? *Could* they do so, if they wanted to? Those are hard questions, not readily answerable by anybody. Throughout the industrial decades, people have become increasingly and more numerously ignorant of the issues of land use, of food, clothing, and shelter. What would they do, and what *could* they do, if they were forced by war or some other calamity to live from their home landscapes?

It is a fact, well attested but little noticed, that our extensive, mobile, highly centralized system of industrial agriculture is extremely vulnerable to acts of terrorism. It will be hard to protect an agriculture of genetically impoverished monocultures that is entirely dependent on cheap petroleum and long-distance transportation. We know too that the great corporations, which grow and act so far beyond the restraint of "the natural affections of the human mind," are vulnerable to the natural depravities of the human mind, such as greed, arrogance, and fraud.

The agricultural industrialists like to say that their agrarian opponents are merely sentimental defenders of ways of farming that are hopelessly old-fashioned, justly dying out. Or they say that their opponents are the victims, as Richard Lewontin put it, of "a false nostalgia for a way of life that never existed."[6] But these are not criticisms. They are insults.

For agrarians, the correct response is to stand confidently on our fundamental premise, which is both democratic and ecological: The land is a gift of immeasurable value. If it is a gift, then it is a gift to all the living in all time. To withhold it from some is finally to destroy it for all. For a few powerful people to own or control it all, or decide its fate, is wrong.

From that premise we go directly to the question that begins the agrarian agenda and is the discipline of all agrarian practice: What is the best way to use land? Agrarians know that this question necessarily has many answers, not just one. We are not asking what is the best way to farm everywhere in the world, or everywhere in the United States, or everywhere in Kentucky or Iowa. We are asking what is the best way to farm in each one of the world's numberless places, as defined by topography, soil type, climate, ecology, history, culture, and local need. And we know that the standard cannot be determined only by market demand or productivity or profitability or technological capability, or by any other single measure, however important it may be. The agrarian standard, inescapably, is local adaptation, which requires bringing local nature, local people, local economy, and local culture into a practical and enduring harmony.

Notes

An earlier version of this essay was published in the twentieth-anniversary issue of *Orion* 21, no. 3 (summer 2002).

1. Vandana Shiva, *Stolen Harvest* (South End Press, 2000), 117.
2. Gerard Manley Hopkins, "God's Grandeur."
3. Virgil, *The Georgics*, trans. L. P. Wilkinson (Penguin Books, 1982), 128.
4. Letter to Rev. James Madison, Oct. 28, 1785.
5. Ibid.
6. Richard Lewontin, "Genes in the Food!" *The New York Review of Books*, June 21, 2001, p. 84.

2

The Resettling of America

Brian Donahue

"Kentucky is not for sale!" declared an agrarian and environmentalist during a discussion of land protection, defying those who would invade his homeland to exploit it for profit. A fine rallying cry—but unfortunately, Kentucky is for sale. Most of the American countryside is privately held, and in America, private land is always for sale. American agrarianism was built around dispersed private ownership of farmland for compelling historical reasons, but it is time to ask whether this can provide a secure foundation for agrarian values in an industrial, market economy. I think the answer is plainly not. Private ownership will continue to have its place, but agrarianism will find its most enduring expression, and its only protection, as commons.

The number of commercial-scale family farms has now been declining for many decades. Private farms have grown to great size, and are often heavily mortgaged—the American heartland is sliding toward corporate ownership. This cannot be the basis for an agrarian culture, which requires that the land be owned by those who live on it in sufficient strength to form communities.[1] Since rural America is now thoroughly unsettled, agrarians must be thinking about how it could be practically resettled, beyond simply waiting for the long-anticipated collapse of industrial society. Who will move to the country, and how could the land be owned and farmed at an agrarian scale?

Meanwhile, other parts of rural America *are* being resettled—not by ignorant farmers, but by "knowledge workers." Fluid communications and transportation make more and more of the countryside attractive and attainable for the affluent. So while farmland is being consolidated into larger ownerships on the production side, it is being fragmented into smaller residential ownerships on the consumption side. The repossession of the countryside by urban Americans is a daunting challenge, but also a great opportunity because it suggests who *might* own an attractive, working rural landscape: not small farmers alone, but villagers. The task before us is to transform suburban sprawl into agrarian village settlement.

I think this can be done. It won't be easy, and it will not succeed before much more land has been lost, but it must be attempted. It is not an impossible task because many people moving to rural areas have motives that, while not precisely agrarian, do have agrarian roots and agrarian possibilities. It is a necessary task because one way or another these folks are going to end up forming most of the rural population, and owning a substantial part of the countryside.

Most suburbanites are not agrarians, of course—their romanticized vision of rural life has been called pastoral, or arcadian. They want to live within a quaintly farmed landscape, but few want to be farmers. They want to live within small, neighborly communities, but they need to drive long distances to work and to shop and have little time left for civic affairs or neighborliness. We are all aware of the destructive contradictions within the suburban impulse, what Leo Marx called sentimental pastoralism—the wish to live in rural harmony by means of industrial exploitation.[2] The question is how to work from this romantic impulse (which simply consumes a pre-existing agrarian landscape) toward a workable agrarian reality.

Somehow, agrarianism will have to flourish in a countryside inhabited mostly by people who are not farmers. How can we get these suburbanites to appreciate and live by agrarian values? That's the nub of the problem. Can we envision agrarian communities where the inhabitants work the land to widely varying degrees (some more, most less), but where all feel vitally connected to the land and its care by complex ties of use and ownership?

First, let me lay out a set of agrarian values—these will be familiar enough to readers of *The Unsettling of America*. The core value, of course, is care of the land. What Wendell Berry calls "kindly use" must be sustainable, not erode soil or degrade waterways. It must not flush excess nutrients or persistent organic pollutants into the environment. And the use of land should have a broad ecological margin—that is, it must allow for the persistence of biodiversity by providing adequate natural (or quasi-natural) habitats such as forests, grasslands, and wetlands. Land that is well used both maintains its productivity and provides an array of ecological benefits.

Well-tended land offers a second value beyond ecological integrity, which is beauty. Open fields and well-built farm buildings, backed by forest or prairie, evoke a powerful feeling of natural harmony in many Americans. The attraction of landscapes that artfully combine nature and culture is deeply agrarian, ancient. This pastoral ideal has acquired a larger border of *wildness* in America during the past century, but that is a gain, as long as it isn't allowed to obscure the entire picture. People who are not farmers will pay to live in diverse rural landscapes. They can learn to pay to keep them alive, as well.

Another fundamental agrarian value is, obviously, the provision of good food. Our farms should be growing, and people should be eating, the most delicious, unadulterated food. It should be, and can be, not only abundant and reasonably priced, but free of pathogens and poisons, and full of flavor and nutrients—which we now understand means, among other things, the right *kind* of fat. If being an affluent society doesn't mean eating well, I can't imagine what it does mean. And of course the same goes for enjoying natural fibers, wood, and other products of the land.

To go with good food, we must have good work. If we want to eat well and be healthy, we need to work hard. This is a natural requirement of our bodies, and again, it's telling that Americans have managed to turn such a wholesome equation into an insoluble pathology—highly profitable to the purveyors of diet and exercise nostrums. We have divided a solution neatly into two problems, as Berry once said about allowing manure to become a pollutant.[3] Agrarian communities must provide many opportunities for physical exercise, contact with nature, and satisfying work. Most people living in rural areas will not earn their living primarily as farmers, but they can participate regularly in agricultural tasks and in other outdoor activities, such as walking and cycling. The work of farming, while demanding, is for the most part pleasant and satisfying. What can be deadening about working the land is not the dirt or sweat, but the economic strain under which it has too often been practiced. So another agrarian value is that outdoor work ought to be enjoyable and widely enjoyed.

Land, beauty, food, work—we can add a fifth agrarian value, that of community. Agrarian communities should be diverse, not segregated by occupation and income as the present suburbs increasingly are. Very few of those who work in my suburban town—the teachers, policemen, tradesmen, landscapers—could ever afford to live there. Those folks live in other suburbs with smaller houses and less open space. Agrarian communities, I hope, would foster much more social and economic interaction among neighbors than we see today in the suburbs. We would also expect a large measure of local democratic control of land use, road construction, housing, and schools. And we would look for a strong "sense of place," a sense of continuity with local history. These communitarian values are not exclusively agrarian, but we would certainly wish to see them flourish in the countryside.

American agrarians have hoped and believed that these characteristics—well-tended land, good food, honest work, beauty, neighborliness—would naturally arise in a country inhabited by independent yeoman farmers. But to flourish, these environmental and community values must coexist with the powerful claims of private property rights and individual profit-maximizing economic behavior. They must somehow be nourished by the market, without succumbing to its narrow logic. So far, this has failed miserably. As a historian I can confirm what Berry argued twenty-five years ago: environmental and community values have not flourished in our agrarian past. They have always been present, but have almost always been submerged by the unkindly use of land and people for short-term profit.

And so there is one bedrock agrarian value that is off my list: the necessity of widely dispersed private ownership of land by independent yeoman farmers to the moral and political health of the democratic republic—the fundamental Jeffersonian ideal. I've excluded it because I don't believe it—I can't believe it and look at the world with any hope. Also, it is contrary to my own experience—the largest piece of private property I've ever owned was a 1963 Jeep pickup truck. So I'm not convinced private landownership is absolutely necessary to agrarian sensibility, or practice. As I look at how the agrarian ideal based on that Jeffersonian principle has fared in American history, at where we are today as a result, and at where we'd like to go, I can only conclude that the ideal of the small yeoman freeholder needs to be greatly overhauled.

We are talking about achieving a decent agrarian future, not recovering an idyllic agrarian past. If the values listed above are the ones we would have the country live by, then we are not going back to anything that ever generally existed in our history. It is essential that we agrarians acknowledge that, as a rule, agrarian life has been pretty bad. In fact, this should not be an *admission* on our part, it should be an *indictment*. Whenever some ag economist asks, "You want to go back to small farming?" and proceeds to describe a life of grinding toil and poverty, we should respond by saying, "You're darn right farming has been hard, and here's the reason why: farming has been hard for most of human history primarily because farmers have been ensnared in political and economic systems designed to extract what they produce, and leave them barely enough to survive."

We agrarians can't be taken seriously unless we begin with the premise that life has been brutally hard for most farm people. Pre-industrial farming was of course an enormous physical labor, marked by low yields, undernourishment, and the threat of disease and famine. It also had its joys and rewards, but they were hard won and fleeting. Advocating agrarianism, and criticizing what industrialism has done to the countryside and to rural culture, does *not* mean advocating a "return" to the conditions of pre-industrial life. Modern science and technology do offer (some) advantages that most agrarians would not only like to keep, but to make more widely available to their fellows around the world. You wouldn't think it would be necessary to affirm this, but I do advise it—otherwise you will quickly find yourself on the defensive, being accused of endorsing the return of bubonic plague.

But the main reason agrarian life was often desperately impoverished was because farmers were being systematically robbed. American agrarian culture came largely (although not exclusively) from Europe, where our peasant ancestors saw close to half of their production directly expropriated by a protection racket run by feudal warlords. Other agrarian cultures—that is, most human societies for the past six or eight thousand years—have rested on similarly downtrodden foundations. American agrarianism developed, by contrast, in a capitalist market economy—under which the crop is frequently worth *less* than what it costs to produce. This means (if I'm doing my math right) that 100 percent of farmers' production is being expropriated, along with some of their other wealth besides—usually income from another job, or the investment in land with which they started. Such agriculture is indeed an economic miracle, but not for farmers.

American soil is rich and abundant. Efficient tools and sustainable methods have been widely available for most of the two centuries of our nationhood. It seems clear that agrarian life here has been hard and unrewarding less because that is the nature of farming, and more because our economic system has made it impossible for most farmers to thrive. The market economy has consistently encouraged and rewarded farming that is exploitative of land and people, and has steadily driven farmers off the land. As it has operated in America, the market has systematically undercut all other agrarian values: care for the land, and healthy family and community life.

It was not supposed to be this way. Those fundamental agrarian aspirations are real. They are not a "myth," ascribed to agrarian life only by naive romantics who were never real farmers themselves. Read the nineteenth-century farm journals, whose subscribers were mostly ordinary farmers. Agrarian values have always been deeply felt by rural Americans, and that remains true today. The problem is that they have almost always been subordinate to "market reality." The distinction Wendell Berry drew in *The Unsettling of America*, between nurturing and exploitative tendencies in our history, is absolutely crucial. The point is not to argue about which is the true description of agrarian life, but to realize that the two tendencies have always existed in tension—but with the exploitative strain almost always the stronger, confounding the nurturing strain even while feeding on it.

American farmers have believed—and most continue to believe, in spite of overwhelming evidence to the contrary—that participating enthusiastically in the market economy would serve agrarian values. They have believed that by owning their own land, working hard and calling their own shots, adopting the latest technologies and competing to grow crops efficiently, they would achieve a prosperous life on the land. But in almost all regions and at almost all times, whole-hearted engagement with the market has led farm families to narrowly specialized, extractive practices, to mounting debt, and either to outright failure or to deciding in the next generation to sell out and try something else—leaving the most aggressive true believers among their neighbors to carry on for another round. Far more often than not, private ownership of land in rural America has not meant an abiding commitment to community and place. It has been a means to accumulate wealth and eventually cash out, provided that the value of the land appreciated faster than the money borrowed against it. Again, the sense of belonging to a place is usually present and deeply valued, but almost always borne down and finally abandoned.

Agrarian values should constitute a system of ethics. Ethics are restraints on immediate self-interest that are held to be ultimately for the common good. They are internalized by individuals, affirmed by cultures, and enforced by laws. It is instructive to look at examples in our history where values of land and community *have* been able to restrain the drive for private profit. The best known—the one we all love to cite—is of course the Amish, among whom many decisions about how work is to be performed are made by the congregation. Limits are agreed to that are judged to be in the lasting interest of the community. Another example of agrarian stability and sustainable husbandry was the colonial New England town, where it was commonplace for a family to work the same soil continuously for generations—not just two or three generations but five, six, seven—lasting well into the nineteenth century. Like the Amish, these communities were founded with a powerful communal religious purpose. That faded over time, but not before establishing a cultural expectation of continuity within the same community and, for at least part of the family, on the same land.[4]

For agrarian values to prevail, either they must be stronger than acquisitive values, or the penetration of the market must be weak. In colonial New England, a diversified, complex approach to farming was maintained because (unlike most American colonies) farmers had few exportable commodities and were on the margins of the market. Most of what they lived on they had to produce locally, and to conserve—they could not, for example, sacrifice all their woodland to maximize their pasture, or they would freeze. When they broke free of these ecological and community restraints in the nineteenth century, developed specialties, and became fully commercial Yankee farmers, they rapidly degraded the land. Most of them also went broke—the "quiet desperation" they felt was often of imminent foreclosure.

Henry Thoreau had it right when he accused his farming neighbors of forgetting the sacredness of their calling, writing that "by avarice and selfishness, and a grovelling habit . . . of regarding the soil as property, or the means of acquiring property chiefly, the landscape is deformed, husbandry is degraded with us, and the farmer lives the meanest of lives." Historians have examined many American farm regions in many periods, and have overwhelmingly confirmed that our agriculture has been driven mainly by rapid extraction of natural capital to supply distant markets. This has been judged by many to be simply "economically rational." It was true of most American colonies before independence, it was quintessentially true of the cotton South (which was the main engine of growth in the American economy before the Civil War), it was largely also true of Free Soil grain farmers of the Midwest, and of course it was wildly true of the cowboy frontier. It has remained true of twentieth-century farming, from the Dust Bowl to the livestock confinement facilities of today—although outright soil mining and slavery have been to some extent replaced by more insidious forms of natural and human degradation. In all these times and places farmers may have wished to form stable, attractive agrarian communities, and they may have worked at it—in particular, the women worked at it. But this was secondary to the main enterprise, which was the pursuit—unsuccessful for most—of individual profit.[5]

It is easy to see why smallholders and tenants escaping feudalism in Europe embraced the opportunity to own land in America and eagerly took advantage of the chance to sell the products of their own labor and raise their standard of living above bare subsistence. It is easy to understand why the American agrarian ideal formed around widespread private ownership of land, and why older systems of landholding that had included some measure of common rights over part of the land were allowed to wither. Such community restraints had formed under conditions of resource scarcity. But there was *so* much land in America (once the original inhabitants had been swept aside), and few immigrants came to this country with communal aspirations as abiding as the New England Puritans, let alone the Amish. America was truly the "best poor man's country," and the right to buy and sell both the land's commodities and the land itself—private property and the free market—became central to our agrarian ideology.[6]

As the agrarian ideal was taking shape in the early Republic, no one could have foreseen that industrial capitalism would grow as spectacularly as it has—even Alexander Hamilton himself couldn't have sanely predicted what we see today. It was perfectly reasonable to suppose, as Jefferson did, that manufacturing would remain subsidiary to a largely agrarian economy—that presumption only looks naive in hindsight, from the far side of a staggering transformation. The idea that farmers would produce first a subsistence for their families and for local exchange, thus rendering themselves economically and politically secure and independent, and only secondarily a surplus for wider markets in order to procure some manufactured and imported goods, made eminent sense. It is simply astonishing how rapidly the abundance and cheapness of fertile soil and rich timber, new tools of cultivation and transportation, rapidly expanding urban and industrial markets in Europe and America, and the sheer exuberance of American acquisitive drive combined to make agricultural (un)settlement such a violently extractive enterprise. The agrarian ideal did not die, but it was submerged and rendered impotent long before the Civil War.

Basing the agrarian ideal on widespread private ownership of land made sense two centuries ago. But now, in retrospect, isn't it clear that in a global market economy dominated by industrial capitalism this is just not an adequate basis for agrarian values of patient devotion to land and community? How long are we going to keep hoping that somehow the temptations of the market can be resisted by virtuous private landowners? But upon what else can we base an agrarian culture?

Clearly, the market economy and private landownership are not going away. They are much too deeply ingrained in our culture, and we will have to deal with them. Besides, they do have their strengths as means to efficiently allocate resources, to provide incentive to generate wealth, and to care responsibly for land, if they can be properly circumscribed and controlled. And we have learned beyond any reasonable doubt that fully collectivized systems are a complete disaster. So the issue today is how to collectively modify and restrain the market, and how to balance private property rights with the common interest in the health of the land.

Looking at the market first, it is clear that we need to keep improving the rules governing the economy to make it more environmentally sound and socially just. This is a matter of curtailing so called externalities—fossil fuel dependence, greenhouse gases, livestock wastes in streams, dead zones in coastal waters, exploitation of migrant laborers, you name it—throughout the industrial economy. Redressing the artificial cheapness of industrial agriculture makes sustainable farming more viable. We all realize that bringing the economy within what *we* consider moral bounds means an ongoing political struggle against corporate power and popular culture. We've made some important gains in the past twenty-five years, but have probably lost more ground than we've made. Key to any success will be continuing to build a constituency that is not only alarmed by the dangers of current industrial practices, but also drawn by practical alternatives.[7]

On that alternative front, we all agree that we do best by making direct connections between farmers and consumers. Successful agrarian marketing requires short-circuiting mainstream distribution channels by means of farmstands, pick-your-own, community supported agriculture (CSAs), and various forms of cooperative organization. Again, all these are familiar approaches, but they have worked well and need to be continued. Our strongest card is appealing to consumers with products that are not only better tasting, but also better for you. The dramatic growth of organic produce in the past quarter century is encouraging, and many of us hope that pastured meat, dairy, and eggs are now poised for a similar takeoff. Grass-based livestock production could have a much more profound impact on the agricultural landscape than a few acres of organic strawberries. These are all good ways to bring more of the food dollar home to the deserving agrarian. Building this economic constituency of consumers goes hand in hand with building a political constituency of citizens willing to curtail the excesses of the industrial economy.

But these market reforms, necessary though they may be, will not be sufficient. One obvious problem is that once we build a new market we can be sure it will quickly be co-opted by larger competitors with more economic clout. They can always bring very similar products to market at a lower price, and if need be they can produce equally delectable homegrown stories to promote them. We see this happening right now with industrial-scale growers and marketers of organic produce. The same will soon happen with grass-fed meat, if we make a showing with that. There isn't much we can do about this except to stay diversified and levelheaded, and to connect with eaters as directly and honestly as we can. If our greatest advantage is connecting with consumers, it helps to have many of those consumers living close at hand. Farmers' markets and websites have their place, but the best direct market is neighbors—people who can visit the farm and appreciate it for more than just its produce.

But there's the rub. What if we did reform the framework of agriculture economics so that small farmers *could* make a decent, reliable living? By itself, that *still* wouldn't keep enough farmers on the land. As long as farmland is privately owned, and freely bought and sold, it is entirely predictable that more and more of it will pass out of farmers' hands. In our industrial economy, or in any environmentally improved version of that economy imaginable, urbanites who want to move to rural areas will be able to outbid farmers for land. Why, these are the very educated and sympathetic consumers we are counting on to support our small farm enterprises! Do we expect them *all* to stay in the cities? Inexorably, even prosperous small farmers will sell out to residential development—*if* their land is their primary asset, and *if* it is worth more growing houses than pasture-farrowed pigs, and *if* we have no means of protecting it. If we are talking, as agrarians must be, about the movement of a substantial part of our population to the countryside, we have to address the issue of how the land is to be owned. If we leave it to the market, we will lose it in the very places where direct marketing is most successful.

Protecting land means moving a significant part of the agrarian landscape into some degree of control by the community. There are several complementary ways to accomplish this. The first is regulation that conveys no ownership per se, but does restrict an owner's right to alter some aspect of his property—wetlands protection is the best-known example.[8] At what point an owner should be compensated for being forbidden to do things with her land that harm the community is of course a matter of great debate. A stronger common interest is conveyed by an easement, by which the community acquires essentially the underlying agrarian or environmental value of the land. The private owner retains the right to occupy and improve the property within agreed limits, and the right to sell or bequeath it. Finally, the community can acquire outright commons in fee simple. As much as possible, in my view, all such common property rights should be held at the local level, under direct democratic control.

There are many ways for a community to acquire common rights in land; to guide development so that it is clustered and the working landscape retained. The work is expensive and slow. Nevertheless, in my part of the world we are hard at it. Connecticut has set a goal of protecting one-quarter of its land. The Society for the Protection of New Hampshire Forests—a leading conservation organization in that state—has also announced an initiative to support every community in conserving at least 25 percent of its land.[9] This is ambitious and laudable, but I live in a suburb in which about 25 percent of the land has been protected and the rest developed, and I can assure you it isn't enough. As an agrarian you might admire some of the things we have done in my town, but you wouldn't want to live there. To have functioning agrarian communities within a healthy natural landscape, we need not a quarter but much more than half of the land to remain undeveloped.

As agrarians, we are advocating both that rural land remain open and working, *and* that people move back to these places. This must involve a fundamental change in how we value and own land. We will need to build a constituency that wishes to inhabit such a landscape—to live within it and to somehow maintain it in that form, intact, rather than subdivided into spacious suburban lots. That is, people moving to rural communities should be buying a small property to call their own, but also a share in the much larger surrounding countryside. How can we accomplish such a transformation of landowning aspirations? One important way to nurture that development—to raise agrarians—is to use part of the protected land in every community for an educational farm.

We have begun this in my suburb of Weston, Massachusetts, and I see it happening more and more in our region and beyond. People are demanding more than mere emptiness of their "open space," and the realization of these new benefits feeds the drive to protect more land. Our town owns about two thousand acres of conservation land, most of it forest but some farmland, and we contract with a nonprofit community farming organization called "Land's Sake" to care for it. Land's Sake maintains trails, mows fields, harvests firewood and timber, makes maple syrup, and grows some twenty-five acres of organic flowers, fruits, and vegetables. The organization has a professional staff but also employs local young people in all of its enterprises, and runs some other educational activities as well. Because we are only ten miles from Boston, the farm is a commercial success.[10]

The agrarian virtues of this are pretty obvious. It takes suburban open space, which would otherwise languish as "viewscape," and turns it back into functioning farmland. More important, it involves young people in learning something about how food and timber are grown. Some have gone on to become farmers and foresters themselves, and they are good farmers, too. In the quarter-century I have been involved with alternative agriculture in New England I have watched the level of competence steadily rise. Here is the breeding ground for the next generation of practicing agrarians. But community farms educate a much larger pool of agrarian supporters as well. The pick-your-own fruits and flowers and the miles of walking trails in the woods draw thousands of urban and suburban residents out into nature on a more casual basis. But as they walk the trails in Weston they may come upon dozens of cords of stacked firewood, and realize they are not visiting a mere "nature preserve." Community farms are ideal vehicles for introducing agrarian values to a suburban population.

Community farms could be a crucial part of making existing suburbs and cities more green, livable places where people will want to stay. It is important to remember that even as the agrarian ideal has been ill-served in America, the urban ideal has been similarly betrayed, and must also be mended. But given that some continued migration to the countryside is desirable and apparently inevitable, community farms are particularly important in rural places where urbanites are just beginning to arrive and we are struggling to keep a large part of the landscape open and working. Wherever land is being protected, energetic young agrarians should be putting together educational farms, or working to organize cooperative stewardship of a few thousand acres of local forest, and getting kids out there on it. Community farms will not replace private farms, but they can play a critical role wherever suburbanization is underway. They are the schools of agrarianism.

The role of community farming will vary across the landscape. Let us imagine a conceptual continuum of suburban market gardens, rural mixed farms, and forest. If a large part of the nation's fruit and vegetable production were returned to the fringes of cities, where it belongs, community farms could serve as a leading provider. Such land simply must be protected by common ownership: public access is highly desirable, and market gardening provides a great opportunity for healthy hand labor. So in denser suburbs a large part of our fruits and vegetables might be grown on some variety of community farm. In rural areas, integrated husbandry of grains, hay, pasture, and livestock seems more the province of small private family farms—but the land itself must be protected. Broader community involvement might be limited mostly to holding easements and to providing young people for summer employment (because farm families are not as large as they once were). Still, one educational farm in each community could serve to deliberately engage more non-farm children in learning a range of agricultural skills. A large part of the nation's forest might be commonly owned at the local level. Most of the forest work would no doubt best be done by highly skilled private contractors, but again, there would be many opportunities for seasonal employment of young people in the woods, too. So, I can see a major role for common ownership and direct community involvement at the market garden and forest ends of the spectrum, and a more limited role for commons in the agrarian middle part of the landscape.

However they are organized, agrarian communities that are populated mainly by non-farmers must have ample means for inhabitants to engage with the land. It is especially important that young people have the chance to do outdoor work as they are going through their high school and college years. And it seems to me that all the inhabitants of such communities should share some ownership of the land, partly through easements and partly as outright commons. In these ways a major part of the population in rural and semi-rural places will be invested in the agrarian landscape and participate actively in agrarian culture. Without this, we will be asking people to make a substantial commitment to maintaining an agrarian environment as merely enlightened observers and consumers, and I don't think they will.

This suggests a new kind of agrarian community. It is similar to the Jeffersonian model in that it envisions a large part of the population engaged with the land, spending at least part of their time laboring in the bosom of the earth and enjoying all the benefits that arise from that. It is similar in that it would be founded on local democratic control of land and community affairs—an idea Jefferson borrowed from the New England township. But it is profoundly different in that it would move away from private ownership of land (although it would certainly not do away with all forms of private ownership) and toward more common rights in land.

A stable agrarian countryside cannot be founded solely on private farmers competing against one another to sell their produce for the lowest price in the marketplace. We cannot have an agrarian society on that basis. Farming is not primarily about growing the cheapest food—that is a dangerous perversion. Farming really must be, first and foremost, a way of life—that has always been the agrarian message. Even if we look at the handful of beleaguered family farmers surviving today and ask why they do it, the only possible answer is because they love farming. They are working two outside jobs, farming at night, and losing money; or they are locked into some usurious contract with a heartless corporation, and *still* they farm. Their agrarian values have been prostituted to the market and so have become the instrument of their own serfdom, and deliver profits only to others.

Agriculture based on agrarian values will require that those who farm do so to fulfill their basic goal of a healthy family life on the land. It will thus require that most farm families have steady off-farm income (as most do today), so that their way of life is not at risk of being lost whenever the market turns sour. It will probably require, in other words, many of what Gene Logsdon calls cottage farmers.[11] I think it will also require many variations of what I am calling community farming—ways of getting most members of the community, especially young people, sufficiently engaged with the land to love and understand it. The object must be to insulate farming from raw market forces so that everyone who farms is doing so primarily for the love of it, and no farmer can be so easily tempted or trapped into destructive ways, or into selling land for some lower and worse use.

Those who travel through rural America have often driven for mile after mile through verdant farmland and seen absolutely nobody out working. Where are the people? They really must be farming at night. Then we reach the beltways surrounding our cities and see tract housing going up at furious pace, often on prime farmland. Our lovely, supremely fertile heartland is now next to deserted, while our inhabited places are nearly uncultivated, except as lawns. Wouldn't it be nice if all of that eerily unsettled rural countryside were instead dense with diversified one-hundred-acre farmsteads, with their grain and hay rotations, livestock, and pastures embedded in a landscape of protected forest, wetland, and prairie? We know that such farming would be at least as productive (per acre) as our present industrialized agriculture, and very nearly as efficient even by the narrowest of economic measures. But we also know it is inconceivable under present market conditions.

For that fantasy to become real we *also* have to visualize each township with its village of one or two thousand people concentrated onto a square mile or so, supporting the surrounding small farms. Imagine such villages across the heartland, one every six miles. That would provide not only something of a direct market, but enough people to sustain a local retail and service economy, to provide ample opportunity for off-farm employment, to maintain churches, schools, and arts. We certainly now have the communications technology so that non-farming people and farmers alike can live in this way without feeling unduly isolated from the mainstream. We can be reasonably sure that enough of them will want to do so to spread over a large part of the countryside in the next century. We want them to have an agrarian sensibility, agrarian values, and above all, agrarian engagement. They will have to have a real stake in the countryside that is part of their community. They will have to walk in it, play in it, do some work in it, and *own* its agrarian-ness—otherwise, they will simply consume it.

The task before agrarians today is to make more agrarians. We cannot protect the countryside by purely defensive measures. There are not enough of us left, and we tend to be poor. Someone is going to have to buy back the land, so we had better turn to those with the cash and the inclination to do it, and make them ours. As Wendell Berry has told us, we have been unsettling America from the start. Now we must learn how to resettle it—or, I should say, to simply begin settling it at last.

NOTES

1. See Marty Strange, *Family Farming: A New Economic Vision* (Lincoln, Nebraska: University of Nebraska Press, 1988).

2. Leo Marx, *The Machine in the Garden: Technology and the Pastoral Ideal in America* (New York: Oxford University Press, 1964).

3. Wendell Berry, *The Unsettling of America: Culture and Agriculture* (San Francisco: Sierra Club Books, 1977), 62.

4. On the Amish see Berry, *Unsettling*, 210–17. On colonial New England see Brian Donahue, *The Great Meadow: The Nature of Husbandry in Colonial Concord, Massachusetts* (Yale University Press, forthcoming 2004).

5. Henry D. Thoreau, *Walden* (Princeton, New Jersey: Princeton University Press, 1973), 165. For a sampling of regional histories dealing with the rural life, land, and the market see Albert E. Cowdrey, *This Land, This South: An Environmental History* (Lexington: University Press of Kentucky, 1983); William Cronon, *Nature's Metropolis: Chicago and the Great West* (New York: W.W. Norton, 1990); Donald Worster, *Dust Bowl: The Southern Plains in the 1930s* (New York: Oxford University Press, 1979); Donald Worster, "Cowboy Ecology," in *Under Western Skies: Nature and History in the American West* (New York: Oxford University Press, 1992) 34–52; Joseph V. Hickey, *Ghost Settlement on the Prairie: A Biography of Thurman, Kansas*, (Lawrence: University Press of Kansas, 1995).

6. James Lemon, *The Best Poor Man's Country: A Geographical Study of Early Southeastern Pennsylvania* (Baltimore: Johns Hopkins Press, 1972)

7. Herman E. Daly and John B. Cobb Jr., *For the Common Good: Redirecting the Economy Toward Community, the Environment, and a Sustainable Future* (Boston: Beacon Press, 1994).

8. See Eric T. Freyfogle's essay, "Private Property Rights in Land: An Agrarian View," in this volume.

9. Society for the Protection of New Hampshire Forests, *New Hampshire Everlasting: An Initiative to Conserve Our Quality-of-Life* (Concord, N.H.: SPNHF, 2001).

10. Land's Sake is detailed in Brian Donahue, *Reclaiming the Commons: Community Farms and Forests in a New England Town* (New Haven: Yale University Press, 1999).

11. Gene Logsdon, *At Nature's Pace: Farming and the American Dream* (New York: Pantheon, 1994).

3

The Mind-Set of Agrarianism . . . New and Old

Maurice Telleen

I don't know quite how to skin this cat called agrarianism. A name like "agrarianism" seems to suggest some widely understood, well-defined sort of movement that maybe just needs a new pair of trousers once in a while to stay in style.

Well, it isn't widely understood in today's world, and as for it being a "movement," I can hope that it isn't. The "ism" bothers me. The track record of people and groups with sure and certain solutions is not pretty. Movements are the natural habitat of true believers. I offer globalization and free trade as the leading current models. Any distrust of that pair is, in some quarters, almost considered evidence of feeblemindedness or terminal backwardness.

Whatever agrarianism is, it is too important to be a mere movement. Movements come and go. Its stepchild and successor, agribusiness, on the other hand, comes with splendid qualifications as a movement, also as an avalanche and a bulldozer. Interestingly enough, agribusiness is very comfortable with the dogmas of free trade, globalization, and multinational corporations that wield incredible power.

Movements, almost by definition, are compelled to be certain or "right." So it is not surprising that they tend to be self-righteous. In addition to being right they are convinced of both their inevitability and their superiority. The latter confers an aura of both practicality and pragmatism on them. Movements leave little room for meaningful dissent. They regard themselves as destiny. Movements are big on tunnel vision. Their tunnel. Their vision. So, let us agrarians give thanks that we are not part of a certified and accredited movement.

Now, having disqualified it as a movement I will next consider it as a religion . . . without the clergy, sacraments, and doctrinal disputes. We'll leave the doctrinal disputes in there. I have no qualms about various degrees of Methodists, Catholics, Muslims, or Jews. We are not fanatics, and since this isn't a movement we don't all have to quack like the AFLAC duck in the commercial. We will even be better off for the differences . . . so long as they don't morph into movements.

I believe religion comes closer to the mark. For agrarianism does have a strong emphasis on personal behavior and its consequences—both long- and short-term—and even eternal life. Eternal life, not as a promise or a reward for being "good" or "saved," but as an inescapable contract that you are stuck with from birth on . . . like it or not, a contract that will continue to play itself out long after you are gone.

Agrarians subscribe generally to that first law of ecology: We can never do just one thing. Action, or inaction, has consequences: both benign and terrible, trivial and important, intended and unintended. We are born into a web of life that both precedes and follows us. Some of it is understood and much of it isn't. But we are each simultaneously part of the picture and one of the painters. Neutrality is not an option. Mindlessness is, but neutrality isn't. That sounds fairly eternal to me. I grant that it has little to do with harps or heaven, but a good bit to do with how and where we use our own time and place on earth. I consider that quasi-religious at the least. It has far more of a moral dimension to it than the free trade, globalization movements previously mentioned. But then, what doesn't?

What I think we can claim for agrarianism is that it is a cultural contract fashioned to work in a specific time and place and that it has great durability and adaptability. Its job description is to function in such a way that it honors and maintains the earth, sustains and perpetuates the community, shelters and benefits the citizens thereof, and respects the commonwealth for what it is: the common wealth. It is neither a warrior society nor an industrial society. It has neither the momentum nor the mobility for that. It is more of a preservationist than a conqueror.

John Maynard Keynes, the great British economist, described pre–World War I, and the root causes of that conflict, in these terms: "A vast and complicated industrial machine, dependent for its working on the equipoise of many factors outside Germany as well as within. The German machine was like a top which to maintain its equilibrium must spin faster and faster."

Does the description of a top, spinning ever faster to maintain its equilibrium, have a familiar ring? We call it the rat race and do a lot of hand wringing about it. But relatively few try to escape it. Jumping off an ever accelerating spinning top is not child's play. Nor is it for sissies.

I suppose the first agrarians were the peasantry. It sure enough wasn't the nobility or the upper crust. Peasantry bears no resemblance whatever to an ever accelerating spinning top.

Agrarianism's natural home is the field, the garden, the stable, the prairie, the forest, the tribe, or the village . . . and the cottage rather than the castle. So it is little wonder that most contemporary Americans are strangers to the term, the concept, and the geography.

Agrarianism has a lot of commonwealth built into it, as in our old understanding of the commons. It served as the cradle of agriculture, before it became agribusiness, and before animal husbandry became animal science. The shift from culture to business was far more than just a linguistic hijacking. Those name changes called forth a whole different mind-set, with a different compass, and in the process both agriculture and animal husbandry have become mere cogs in giant industries. Culture and husbandry have been shoved aside with contempt by these same giant industries that are part and parcel of the movements known as globalization and free trade.

This didn't happen overnight. In some respects the consequences were unintended, or at least unanticipated by the very folks that made them happen. Many of them were beneficial, and remain so to this day. Others were so incremental that they went unnoticed by the "news vendors" for decades. The evening news rarely deals with things incremental. It deals with explosions, not glacial-like changes.

The agrarianism that Gene Logsdon, Wendell Berry, and I, along with many others, grew up with was an intermediate model. It was one of the later chapters in the long story of agriculture adapting to industrialism, that hinge event that took a couple hundred years. It had been tinkered with incrementally from the onset of the industrial revolution, often improved, always subject to change, and for my part I figured that tinkering would continue. I never dreamt, as a kid, that agrarianism (a word I never heard as a child, even as we practiced it) would someday be challenged as obsolete, and even contemptible in some quarters. But during that time agrarianism was slowly losing its natural habitat, which is tough on more than owls and pussy cats.

Eventually the ideas of endless progress and endless growth (also a disease known as cancer) elbowed out the older notions of stability, maintainability, and continuity. Even the notion of what constituted wealth changed. That semantic tinkering from culture to business in one case and from husbandry to science in the other changed the whole social, physical, and economic landscape of both rural and urban America. It rarely made the evening news. Which is not to say that it went unnoticed.

Wendell Berry, for one, noticed. He made his concerns known in the publication of his book *The Unsettling of America,* a book that justly deserves to be honored. I believe Berry hoped that the book would trigger a great debate, a soul searching dialogue on just who in the hell let all the chickens out of the henhouse in the midst of so many foxes, and what the long-term consequences were likely to be.

I might add that he had a partner in that enterprise in the form of the Sierra Club, who first published the book. I would hope that this book might rekindle, or kindle in many cases, the partnership between farmers and conservationists. While our primary concerns are by no means identical, they are certainly kissing cousins. It is time they recognized their commonality and became better acquainted.

In any event, the great national debate did not materialize, notwithstanding continuous debate and dialogue. But for the most part, mainstream agribusiness just shrugged and almost ignored the indictment like a cow or horse uses its tail to switch flies—absentmindedly. When events and the tide are on your side, perhaps you can take liberties like that.

Perhaps the era of great public debates is over. Or maybe the stuff of such debates just doesn't fit into neat little sixty- to ninety-second blurbs around the commercials on the six o'clock news.

Agrarianism has even had some occasional successes, but the fundamental questions drift about unaddressed by mainstream agribusiness. While at the same time the rural areas of this country offer plenty of evidence that there was, and is, cause for an in-depth examination of the shape and future of farming. The fact that there are growing concerns about the future of farming and food indicates that the flies have not gone away either in defeat or shame.

One reason, I believe, for being ignored is that agrarianism isn't just about money. It might get a more respectful hearing if it were. But it is about culture, just as agriculture was about culture. Before it got run into the ditch by agribusiness.

A funny thing about cultures is that they produce people who understand more than they know. Sort of like osmosis. So the old agrarians, to get back to our subject, knew a lot about local soil, local weather, local crops, animal behavior, and each other. They depended on each other. It almost defines that much abused word, provincial. It was very provincial and no doubt carried a load of both inertia and foolishness, along with wisdom.

But whatever the mix, it was rooted to places, communities, continuity, and people whose names and faces you knew. As a matrix, it worked reasonably well. Which is different from claiming that it was idyllic and completely satisfactory. To assert that it was perfect is nostalgia, also known as remembrance through rose-tinted glasses. Nostalgia, except in the tourist trade, is considered a bad disease these days, almost as bad as barter to an IRS agent. But where tourism is considered, it is a tangible asset. Check that out with the folks who run tour buses.

There is a tour bus story that happens to be true about the young lady who, seeing an Amish man out plowing in Lancaster County, Pennsylvania, asked the driver to stop so she could get his picture. So she went down to the field and waited. When he got to the headland she figured he would stop and chat and smile when she said "cheese." Instead he simply turned the horses around and kept on plowing. She was incensed by his insubordination. Can't you just see it? Here she was, "Yoo-hoo, Mr. Amishman, please stop. I want to get your picture."

He, on the other hand, had other priorities and probably thought, "What is that crazy woman yelling about?" and drove on.

She, on the other hand (you need three hands in this business) went to the Chamber of Commerce and complained about his behavior. Probably told them that with employees like that they could expect fewer tourists. It was as though the farmer was a character actor in a theme park who had treated one of the customers rudely, rather than a member of a community doing the work of maintaining a culture.

So where does that leave the term "agrarianism" today? Is it more than simply an anachronism to many, irrelevant to others, or a job in a theme park for someone else? Should agrarianism matter to folks who are not farmers, since it is hard to grow vegetables in an apartment, keep a family cow in the garage, and garden on the roof (unless you have an earth house)? So far, agrarianism has been useful to mainstream agribusiness only as an ancient virtue to invoke at election time, or when a farm bill is before Congress, or in touchy-feely television commercials. In any event, if the future of agrarianism is limited to ruralists who farm and garden, I see the whole concept as little more than a small obstacle in a headlong race to God knows what.

Nearly all of today's success stories in agrarianism are about niches and making it in the margins. I'll cite our own experience in that regard. At one time we had a list of about twenty households who depended on us for locker lambs, a few of them for two a year. They got a good product and we got a premium price, as well as the pleasure of their friendship. Had ten other flock owners around Waverly, Iowa, tried it, or worked at it, there would have been little in it for any of us. A family can patronize only one producer and consume only so many sheep. So most of us would have quit. All of which is sort of a microcosm of modern agribusiness. The great majority of hog producers in my own state have quit.

We also started a magazine called the *Draft Horse Journal* in the mid-1960s. The draft horse, by that time, had been dismissed by agribusiness as a hopeless and useless artifact. Like Egyptian mummies, interesting to a few oddballs no doubt, but of no practical value to a modern funeral parlor or farmer. It had been pronounced dead and buried by platoons and even regiments of economists. The journal flourished and managed to keep us sometimes busier than we cared to be for the rest of our lives, so far. Had three or four others tried it at the same time, we all would have gone broke. And the expectations of the conventional wisdom would have been proven correct.

It wasn't a big enough niche to even attract predators. Publishing giants are looking for bigger meals. It has, however, attracted a couple of start-up competitors. But they are small potato, agrarian-type operations too. And, strangely enough, we all three seem to manage to put beans on our own tables. I think the horse business is better off for it with three different organs, each one playing a little different tune.

So is this the time for a publishing giant to step into the breach and produce one super-duper draft horse magazine by swallowing all of us? If most of the once-great daily papers in this country that have been swallowed up by the giants are any indication, I'd say the answer would be a resounding *No!* There is, of course, such a thing as economy of scale but I'm sure that any of the giants could wreck all three of them in no time at all. Then there would be none. Consider also the case of Fred Kirschenmann, who turned his farming operation in North Dakota into an agrarian showcase. I suspect that, too, owed much of its success to the fact that half of North Dakota's other farmers did not follow him down that same road.

And so it goes all over the country as our rural infrastructure continues to deteriorate, in spite of these small examples of agrarian success in this sea of rural destruction.

The problems are not "going away" like a pinched nerve or the pimples of adolescence. We have much work to do. I can't even imagine what some of it will be. But I am convinced that the mind-set of agrarianism has a validity well beyond food production. And I believe that mind-set can give strength and stability to our whole society. It is, obviously, our duty to share it with the rest of the country.

Since this is a quasi-religion it needs some commandments . . . ten sounds like just the right number. Here they are, free for the taking. If you want to add to, subtract, or modify that's okay. Agrarianism is *not a movement*. It is, in fact, a fairly loose-fitting garment probably most at home in the country, but you can wear it in town, too. This may save this country's bacon. Wouldn't that be neat?

The Ten Commandments of Agrarianism
. . . and commentary on same

1. The earth is our mother, not always forgiving and not a treasure house to plunder. And we all drink out of the same well, but first locally. *Commentary:* As Garret Hardin says in *Living with Limits*, that wonderful book of his, "Every resident of Manhattan, whether he knows it or not, and no matter how crowded he may feel, lives on (or off) more than nine acres of land. Since most of the acreage the average citizen occupies is out of sight and mind, the agricultural geographer Georg Borgstrom suggested in 1961 that we call it 'ghost acreage.' The essential life of an educated urban dweller, from birth to death, is lived out on ghost acreage."[1]
2. Enough is enough. Optimum is often better than maximum. And don't bother to keep up with the Joneses. They probably don't know where they are going, either. *Commentary:* About twenty-five hundred years ago the Greeks inscribed this advice on the temple at Delphi: "Nothing in excess." As Hardin says, "Oxygen is a good thing, but breathe 100% oxygen for a few hours and you are dead." I'll bet on the Greeks, rather than today's cheerleaders in both politics and the advertising racket.
3. Like the physician, first strive to do no harm. *Commentary:* That is harder than it sounds. It does not mean doing absolutely nothing . . . except once in a while.
4. Don't feed the seed corn to the steers or the chickens, and don't bet all your marbles on the same horse. *Commentary:* I don't think I should have to explain everything.
5. Remember that the term "my country" is not just about flags, parades, and anthems sung badly. It is about "places" like Jefferson's Virginia, Adams's New England, Gene Logsdon's Ohio, Wendell Berry's Kentucky, and my Iowa. *Commentary:* When the old Scottish livestock breeder said he was trying to produce beasts that would *do the country*

good, he meant his own neck of the woods (or moors). He was not speaking of high-priced exports to North America.

6. Accept limits with grace. *Commentary:* Limits are not shackles; in fact, many of them are liberating. We are not meant to be "gods" or "Masters of the Universe," and Lord knows we have the track record to prove it.
7. Ecclesiastes had it right. There is an inescapable and necessary order of things, with a time for everything, including billiards and softball. *Commentary:* "Certain seasons require certain kinds of work; there is a breaking season, a planting season, a cultivating season, a laying by time, and a marketing time. This very loose organization is determined by nature, not by man and points to the fundamental difference between the factory and the soil. Industrialism is multiplication. Agrarianism is addition and subtraction. The one, by attempting to reach infinity must become self-destructive; the other by fixing arbitrarily its limits upon nature will stand."[2]
8. This is the big one: Beware of nature-free equations and computer models, especially in the hands of those who talk glibly about "the creation of wealth" and endless growth. They speak only of money. And distrust all the military metaphors that fill our lives. *Commentary:* We are constantly being urged to wage war against drugs, weeds, teen pregnancies, fat people, poverty, snotty noses, and on and on. Final, complete, and unconditional victory is one of our favorite myths. And we have the metaphors to prove it.
9. Be suspicious of the "Gee Whiz" and "Next Big Thing" mentality, not from a distrust of whatever is new (some of it is wonderful) but with the sure and certain knowledge that whatever it is, it too comes with some baggage. *Commentary:* And no matter how wonderful, some of the baggage will have sharp teeth.
10. Distrust the "big picture." Where you live is where you matter most, and arresting those gullies on the slopes of your own hillsides is more important than seeing the Grand Canyon. *Commentary:* I have never seen the Grand Canyon. I suppose I really should make an attempt someday.

Ten. That was enough for Moses and Charlton Heston, so it ought to be enough for us. Agrarianism in today's world must, of necessity, be a fairly loose garment, because so many of us live in places where you can't even pull weeds. But you don't have to have dirt under your fingernails for the mind-set to be useful wherever you live, whether in Manhattan, New York, or Manhattan, Kansas.

Notes

1. Garret Hardin, *Living with Limits* (New York: Oxford University Press, 1993).
2. Andrew Nelson Lytle, "The Hind Tit," in *I'll Take My Stand: The South and the Agrarian Tradition by Twelve Southerners* (Baton Rouge: Louisiana State University Press, 1930).

4

Sustainable Economic Development

Definitions, Principles, Policies

Herman E. Daly

I. Definitions

Exactly what is it that is supposed to be *sustained* in "sustainable" economic development? Two broad answers have been given:

The first answer states that *utility* or happiness should be sustained; that is, the utility of future generations is to be non-declining. People in the future should be at least as well off as those living in the present in terms of the levels of happiness they can experience. "Utility" here refers to average per capita utility of members of a generation.

The second answer states that physical *throughput*, the entropic physical flow from nature's sources through the economy and back to nature's sinks, should be sustained and non-declining. More exactly, the capacity of the ecosystem to sustain energy/food flows over the long term is not to be run down. Natural capital, which can be defined here as the capacity of ecosystems to yield a durable flow of natural resources and a flux of natural services, is to be kept intact. The future should be at least as well off as the present in terms of its access to biophysical resources and services supplied by the ecosystem. "Throughput" here refers to total throughput flow for the community over some time period (i.e., the product of per capita throughput and population).

These are two totally different concepts of sustainability. Utility is a basic concept in standard economics. Throughput is not, in spite of the efforts of Kenneth Boulding and Nicholas Georgescu-Roegen to introduce it. So it is not surprising that the utility definition has been dominant.

Nevertheless, I adopt the throughput definition and reject the utility definition, for two reasons. First, utility is nonmeasurable. Second, and more important, even if utility were measurable it is still not something that we can bequeath to the future. Utility is an experience, not a thing. We cannot bequeath utility or happiness to future generations. We can leave them things, and to a lesser degree knowledge (provided that it is actively pursued and not simply passively received). Whether future generations will make themselves happy or miserable with these gifts is simply not under our control. To define sustainability as a non-declining intergenerational bequest of something that can neither be measured nor bequeathed strikes me as a nonstarter.[1] I hasten to add that I do not think economic theory can get along without the concept of utility. I just think that throughput is a better concept by which to define sustainability.

The throughput approach defines sustainability in terms of something much more measurable and transferable across generations—the capacity to generate an entropic throughput from and back to nature.[2] Moreover this throughput is the metabolic flow by which we live and produce. The economy in its physical dimensions is made up of things—populations of human bodies, livestock, soils, plants, machines, buildings, and artifacts. All these things are what physicists call "dissipative structures" that are maintained against the forces of entropy by a throughput from the environment. An animal can only maintain its life and organizational structure by means of a metabolic flow through a digestive tract that connects to the environment at both ends. So too with all dissipative structures and their aggregate, the human economy. All economies depend on what Wendell Berry has called the "Great Economy," the vast network of patterns and powers in terms of which all of life's necessities and values are parceled out and exchanged.

Economists are very fond of the circular flow vision of the economy, inspired by the circulation of blood discovered by William Harvey in the early seventeenth century, emphasized by the Physiocrats, and reproduced in the first chapter of every economics textbook. Somehow the digestive tract has been less inspirational to economists than the circulatory system. An animal with a circulatory system, but no digestive tract, could it exist, would be a self-enclosed perpetual motion machine. Biologists do not believe in perpetual motion machines. Nor do ecologists, who understand that living beings cannot survive alone but exist from and through their memberships with other organisms and their sustaining habitats. Economists, however, seem dedicated to keeping an open mind on the subject.

Bringing the concept of throughput into the foundations of economic theory does not reduce economics to physics, but it does force the recognition of the constraints of physical/biological laws on economics. Among other things, it forces the recognition that "sustainable" cannot mean "forever," for as scientists tell us, the physical world is temporally finite, and will likely end either in a big cooling or a big crunch. Sustainability is a way of asserting the value of longevity and intergenerational justice, while recognizing mortality and finitude. Sustainable development is not a religion, although some seem to treat it as such. Since large parts of the throughput are nonrenewable resources, the expected lifetime of our economy is much shorter than that of the universe. Sustainability in the sense of longevity requires increasing reliance on the renewable part of the throughput, and a willingness to share the nonrenewable part over many generations.[3] Of course longevity is no good unless life is enjoyable, so we must give the utility definition its due in providing a necessary baseline condition. That said, in what follows I adopt the throughput definition of sustainability and will have nothing more to say about the utility definition.

Having defined "sustainable," let us now tackle "development." Development might more fruitfully be defined as more utility per unit of throughput, and growth defined as more throughput. But since current economic theory lacks the concept of throughput, we tend to define development simply as growth in gross domestic product (GDP), a value index that conflates the effects of changes in throughput and utility.

Defining development in terms of the growth of GDP highlights a significant problem in the way economic accounting occurs. The prices used in calculating this value index are, of course, affected by distributions of wealth and income, by the exclusion of the demands future generations and nonhuman species might have, and by the failure to include other external costs and benefits into prices. Values are thus a reflection of the interests of a limited subset of the Great Economy (primarily the interests of those connected with the management of financial capital), whereas what may be valuable to the poor (clean water and air) and other organisms (healthy habitats) hardly registers at all. It is hard to give a normative meaning to an index constructed with such distorted relative prices.

The hope that the growth increment will go largely to the poor, or at least trickle down, is frequently expressed as a further condition of development. Yet any serious policy of redistribution of GDP from rich to poor is currently rejected as "class warfare" that is likely to slow GDP growth. Furthermore, any recomposition of GDP from private goods toward public goods (available to all, including the poor) is usually rejected as government interference in the free market—even though it is well known that the free market will not produce public goods. We are assured that a rising tide lifts all boats, that the benefits of growth will eventually trickle down to the poor. The key to development is still aggregate growth, and the key to aggregate growth is currently thought to be global economic integration—free trade and free capital mobility. Export-led development is considered the only option. Import substitution is no longer mentioned, except to be immediately dismissed as "discredited."

Will this theory or ideology of "development as global growth" be successful? I doubt it, for two reasons, one having to do with environmental sustainability, the other with social equity.

First, ecological limits are rapidly converting "economic growth" into "uneconomic growth"—i.e., throughput growth that increases costs by more than it increases benefits, thus making us poorer not richer. The macroeconomy is not the Whole—it is Part of a larger Whole, namely the ecosystem. As the macroeconomy grows in its physical dimensions (throughput), it does not grow into the infinite Void. It grows into and encroaches upon the finite ecosystem, thereby incurring an opportunity cost of preempted natural capital and services. These opportunity costs (depletion, pollution, sacrificed ecosystem services) can be, and often are, worth more than the extra production benefits of the throughput growth that caused them. We cannot be absolutely sure because we measure only the benefits, not the costs.[4] We do measure the regrettable defensive expenditures made necessary by the costs, but even those are *added* to GDP rather than subtracted. For example, the medical bills resulting from the treatment of respiratory illness caused by air pollution are added to GDP, rather than netted out as a cost of production of those activities that caused the air pollution.

Second, even if growth entailed no environmental costs, part of what we mean by poverty and welfare is a function of relative rather than absolute income, that is, of social conditions of distributive inequality. Growth cannot possibly increase everyone's *relative* income. Insofar as poverty or welfare is a function of relative income, then growth becomes powerless to affect it.[5] This consideration is more relevant when the growth margin is devoted more to relative wants (as in rich countries) than when devoted more to absolute wants or needs (as in poor countries). But if the policy for combating poverty is *global* growth, then the futility and waste (not to mention injustice) of growth dedicated to satisfying the relative wants of the rich cannot be ignored.

Am I saying that wealth has nothing to do with welfare, and that we should embrace poverty? Not at all! More wealth is surely better than less, up to a point. The issue is, does growth increase net wealth? How do we know that throughput growth, or even GDP growth, is not at the margin increasing *illth,* a term coined by John Ruskin to describe the opposite of wealth (a growth in "bads" rather than "goods"), faster than *wealth*, making us poorer, not richer? Illth accumulates as pollution at the output end of the throughput, and as resource depletion at the input end. Ignoring physical throughput in economic theory thus leads to treating depletion and pollution as "surprising" external costs, if recognized at all. Building the throughput into economic theory as a basic concept allows us to see that illth is necessarily generated along with wealth. When a growing throughput generates illth faster than wealth, as when nuclear power generates higher production and toxic waste disposal costs than electrical energy benefits, then its growth has become uneconomic. Since macroeconomics lacks the concept of throughput, it is to be expected that the concept of "uneconomic growth" will not make sense to macroeconomists.

While growth in rich countries might be uneconomic, growth in poor countries, where GDP consists largely of food, clothing, and shelter, is still very likely to be economic. Food, clothing, and shelter are absolute needs, not self-canceling relative wants for which growth yields no welfare. There is much truth in this, even though poor countries too are quite capable of deluding themselves by counting natural capital consumption (depleting mines, wells, forests, fisheries, and topsoil) as if it were Hicksian income.[6] One might legitimately argue for limiting growth in wealthy countries (where it is becoming uneconomic) in order to concentrate resources on growth in poor countries (where it is still economic).

The current policies of the International Monetary Fund (IMF), the World Trade Organization (WTO), and the World Bank (WB), however, are decidedly *not* for the rich to decrease their *uneconomic* growth to make room for the poor to increase their *economic* growth. The concept of uneconomic growth, which depends on the acknowledgments of physical throughput and illth, remains unrecognized. Rather, the vision of globalization requires the rich to grow rapidly in order to provide markets in which the poor can sell their exports. It is thought that the only option poor countries have is to export to the rich, and to do that they have to accept foreign investment from corporations who know how to produce the high-quality stuff that the rich want. The resulting necessity of repaying these foreign loans reinforces the need to orient the economy toward exporting and exposes the borrowing countries to the uncertainties of volatile international capital flows, exchange rate fluctuations, and unpayable debts, as well as to the rigors of competing with powerful world-class firms.

The whole global economy must grow for this policy to work, because unless the rich countries grow rapidly they will not have the surplus to invest in poor countries, nor the extra income with which to buy the exports of the poor countries.

The inability of macroeconomists to conceive of uneconomic growth is very strange, given that microeconomics is about little else than finding the optimal extent of each micro activity. An optimum, by definition, is a point beyond which further growth is uneconomic. The cardinal rule of microeconomic optimization is to grow only to the point at which marginal cost equals marginal benefit. That has been aptly called the "when to stop" rule—when to stop growing, that is. Macroeconomics has no when to stop rule. GDP is supposed to grow forever.[7] The reason is that the growth of the macroeconomy, currently modeled on the self-enclosed, perpetual motion circulatory system rather than the open and interdependent digestive system, is not thought to encroach on anything and thereby incur any growth-limiting opportunity cost. By contrast the microeconomic parts grow into the rest of the macroeconomy by competing away resources from other microeconomic activities, thereby incurring an opportunity cost. The macroeconomy, however, is thought to grow into the infinite Void, never encroaching on or displacing anything of value. The point to be emphasized is that the macroeconomy too is a Part of a larger finite Whole, namely the ecosystem. The optimal scale of the macroeconomy relative to its containing ecosystem is the critical issue to which macroeconomics has been blind. This blindness to the costs of growth in scale is largely a consequence of ignoring physical throughput and has led to the problem of ecological unsustainability.

II. Growth by Global Integration: Comparative and Absolute Advantage and Related Confusions

Under the current ideology of export-led growth, the last thing poor countries are supposed to do is to produce anything for themselves. Any talk of import substitution is nowadays met by trotting out the abused and misunderstood doctrine of comparative advantage. The logic of comparative advantage is unassailable, given its premises. Unfortunately one of its premises (as emphasized by David Ricardo) is capital immobility between nations. When capital is mobile, as indeed it is, we enter the world of absolute advantage, where, to be sure, there are still global gains from specialization and trade. However, there is no longer any guarantee that *each* country will necessarily benefit from free trade as under comparative advantage. One way out of this difficulty would be to greatly restrict international capital mobility, thereby making the world safe for comparative advantage.[8] The other way out would be to introduce international redistribution of the global gains from trade resulting from absolute advantage. Theoretically the gains from absolute advantage specialization would be even greater than under comparative advantage because we would have removed a constraint to the capitalists' profit maximization, namely the international immobility of capital. But absolute advantage has the political disadvantage that there is no longer any guarantee that free trade will mutually benefit all nations. Which solution does the IMF advocate—comparative advantage vouchsafed by capital immobility, or absolute advantage with redistribution of gains to compensate losers? Neither. They prefer to pretend that there is no contradiction, and call for both comparative advantage–based free trade and free international capital mobility—as if free capital mobility were a logical extension of comparative advantage–based free trade instead of a negation of its premise. This is incoherent.

In an economically integrated world, one with free trade and free capital mobility, and increasingly free, or at least uncontrolled, migration, it is difficult to separate growth for poor countries from growth for rich countries, since national boundaries become economically meaningless. Only by adopting a more nation-based approach to development can we say that growth should continue in some countries but not in others. But the globalizing trio, the IMF, WTO, and WB cannot say this. They can only advocate continual global growth in GDP. The concept of uneconomic growth just does not compute in their vision of the world. Nor does their cosmopolitan ideology recognize the nation as a fundamental unit of community and policy, even though their founding charter defines the IMF and World Bank as a federation of nations. Local self-sufficiency at the sub-national level is a highly desirable goal, but will fly further out of our reach unless we stop global economic integration.

III. Ignoring Throughput in Macroeconomics: GDP and Value Added

As noted, throughput and the scale of the macroeconomy relative to the ecosystem are not familiar concepts in economics. Therefore let us return for a while to the familiar territory of GDP and value added and approach the concept of throughput by this familiar path. Economists define GDP as the sum of all value added by labor and capital in the process of production. Exactly what it is that value is being added to is a question to which little attention is given, since current approaches to valuation do not include natural resources. The gasoline price, for instance, does not factor the gasoline itself into the cost, but rather the labor and capital expended in drilling, pumping, and refining it. The value of the gasoline *in situ* is taken as zero.[9]

Value added is simultaneously created and distributed in the very process of production. Therefore, economists argue that there is no GDP "pie" to be independently distributed according to ethical principles. As Kenneth Boulding put it, instead of a pie, there are only a lot of little "tarts" consisting of the value added by different people or different countries and mindlessly aggregated by statisticians into an abstract "pie" that doesn't really exist as an undivided totality. If one wants to redistribute this imaginary "pie" one should appeal to the generosity of those who baked larger tarts to share with those who baked smaller tarts, not to some invidious notion of equal participation in a fictitious common inheritance.

I have considerable sympathy with this view, as far as it goes. But it leaves out something very important.

In our one-eyed focus on value added we economists have neglected the correlative category, "that to which value is added," namely the throughput. "Value added" by labor and capital has to be added *to something*, and the quality and quantity of that something is important. There is a real and important sense in which the original contribution of nature is indeed a "pie," a pre-existing, systemic totality that we all share as an inheritance. It is not an aggregation of little tarts that we each baked ourselves. Rather it is the seed, soil, sunlight, and rain from which the wheat and apples grew that we then converted into tarts by our labor and capital. The claim for equal access to nature's bequest is not the invidious coveting of what our neighbor produced by her own labor and abstinence. The focus of our demands for income to redistribute to the poor, therefore, should be on the value of the contribution of nature, the original value of the throughput to which further value is added by labor and capital—or, if you like, the value of low entropy added by natural processes to neutral, random, elemental stuff. Failure to acknowledge this original value, as we now know, leads to its exploitation and destruction. Or as Berry puts it, we increasingly see human economies succeeding by the invading and pillaging of the Great Economy.

IV. Ignoring Throughput in Microeconomics: The Production Function

But there is also a flaw in our very understanding of production as a physical process. Neoclassical production functions are at least consistent with the national accountant's definition of GDP as the sum of value added by labor and capital, because they usually depict output as a function of only two inputs, labor and capital. Land, as the geophysical basis on which natural and commercial growth depends, does not appear. In other words, value added by labor and capital in production is added to nothing, not even valueless neutral stuff. But value cannot be added to nothing. Neither can it be added to ashes, dust, rust, and the dissipated heat energy in the oceans and atmosphere (i.e., high-entropy matter/energy). The lower the entropy of the resource (i.e., concentrated ores and mineral deposits) the more capable it is of receiving the imprint of value added by labor and capital. High entropy resists the addition of value. Since human action cannot produce low entropy in net terms we are entirely dependent on nature for this ultimate resource by which we live and produce.[10] Any theory of production that ignores this fundamental dependence on throughput is bound to be seriously misleading.

As an example of how students are systematically misled on this issue I cite a textbook[11] used in the microeconomic theory course at my institution. On page 146 the student is introduced to the concept of production as the conversion of inputs into outputs via a production function. The inputs or factors are listed as capital (K), labor (L), and materials (M)—the inclusion of materials is an unusual and promising feature. We turn the page to page 147 where we now find the production function written symbolically as **q=f(K, L)**. M has disappeared, never to be seen again in the rest of the book. Yet the output referred to in the text's "real world example" of the production process is "wrapped candy bars." Where in the production function are the candy (sugars, corn syrup, cocoa, etc.) and wrapping paper (trees) as inputs?[12] Production functions are often usefully described as technical recipes. But unlike real recipes in real cookbooks we are seldom given a list of ingredients!

And even when neoclassicals do include resources as a generic ingredient it is simply "R" raised to an exponent and multiplied by L and K, also each raised to an exponent. Such a multiplicative form means that R can approach zero if only K and L increase sufficiently. Presumably we could produce a one hundred pound cake with only a pound of sugar, flour, eggs, etc., if only we had enough cooks stirring hard in big pans and baking in a big enough oven!

The problem is that the production process is not accurately described by the mathematics of multiplication. Nothing in the production process is analogous to multiplication.[13] What is going on is *transformation*, a fact that is hard to recognize if throughput is absent. R is that which is being transformed from raw material to finished product and waste (the latter symptomatically is not listed as an output in production functions). R is a flow. K and L are agents of transformation, stocks (or funds) that effect the transformation of input R into output Q—but which are not themselves physically embodied in Q. There can be substitution between K and L, both agents of transformation, and there can be substitution among parts of R (aluminum for copper), both things undergoing transformation. But the relation between agent of transformation (efficient cause) and the material undergoing transformation (material cause) is fundamentally one of complementarity. Efficient cause is far more a complement than a substitute for material cause! This kind of substitution is limited to using a little extra labor or capital to reduce waste of materials in process—a small margin soon exhausted.[14]

Language misleads us into thinking of the production process as multiplicative, since we habitually speak of output as "product" and of inputs as "factors." What could be more natural than to think that we multiply the factors to get the product! That, however, is mathematics, not production! If we recognized the concept of throughput we would speak of "transformation functions," not production functions.

V. Opposite Problems: Non-enclosure of the Scarce and Enclosure of the Non-scarce

Economists have traditionally considered nature to be infinite relative to the economy, and consequently not scarce, and therefore properly priced at zero. But nature is scarce, and becoming more so every day as a result of throughput growth. Efficiency demands that nature's services be priced, as even Soviet central planners eventually discovered. But to whom should this price be paid? From the point of view of efficiency it does not matter who receives the price, as long as it is charged to current users (to expect future generations to absorb all these costs is to raise a host of complex moral questions). But from the point of view of equity it matters a great deal who receives the price for nature's increasingly scarce services. Such payment is the ideal source of funds with which to fight poverty and finance public goods.

Value added belongs to whoever added it. But the original value of that to which further value is added by labor and capital should belong to everyone. Scarcity rents to natural services, nature's value added, should be the focus of redistributive efforts. Rent is by definition a payment in excess of necessary supply price, and from the point of view of market efficiency is the least distorting source of public revenue.

Appeals to the generosity of those who have added much value by their labor and capital are more legitimate as private charity than as a foundation for fairness in public policy. Taxation of value added by labor and capital is certainly legitimate. But it is both more legitimate and less necessary after we have, as much as possible, captured natural resource rents for public revenue.

The above reasoning reflects the basic insight of Henry George (*Progress and Poverty*), extending it from land to natural resources in general. Neo-classical economists have greatly obfuscated this simple insight by their refusal to recognize the productive contribution of nature in providing "that to which value is added." In their defense it could be argued that this is so because in the past economists considered nature to be non-scarce, but now they are beginning to reckon the scarcity of nature and enclose it in the market. Let us be glad of this, and encourage it further.

Although the main problem I am discussing is the non-enclosure of the scarce, an opposite problem (enclosure of the non-scarce) should also be noted. There are some goods that are by nature non-scarce and non-rival and should be freed from illegitimate enclosure by the price system. I refer especially to knowledge. Knowledge, unlike throughput, is not divided in the sharing, but multiplied. There is no opportunity cost to me from sharing knowledge with you. Yes, I would lose the monopoly on my knowledge by sharing it, but we economists have long argued that monopoly is a bad thing because it creates artificial scarcity that is both inefficient and unjust. Once knowledge exists, the opportunity cost of sharing it is zero and its allocative price should be zero. *Consequently, I would urge that international development aid should more and more take the form of freely and actively shared knowledge, and less and less the form of interest-bearing loans.* Sharing knowledge costs little and does not create unpayable debts, and it increases the productivity of the truly scarce factors of production.

Although the proper allocative price of existing knowledge is zero, the cost of production of new knowledge is often greater than zero, sometimes much greater. This of course is the usual justification for intellectual property rights in the form of patent monopolies. Yet the main input to the production of new knowledge is existing knowledge, and keeping the latter artificially expensive will certainly slow down production of the former. This is an area needing much reconsideration. I only mention it here, and signal my skepticism of the usual arguments for patent monopolies, so emphasized recently by the free-trading globalizers under the gratuitous rubric of "trade-related intellectual property rights." As far as I know, James Watson and Francis Crick receive no patent royalties for having unraveled the structure of DNA, arguably the most basic scientific discovery of the twentieth century. Yet people who are tweaking that monumental discovery are getting rich from monopolizing their relatively trivial contributions that could never have been made without the free knowledge supplied by Watson and Crick.

Although the main thrust here is to bring newly scarce and truly rival natural capital and services into the market enclosure, we should not overlook the opposite problem, namely, freeing truly non-rival goods from their artificial enclosure by the market.

VI. Principles and Policies for Sustainable Development

I am not advocating revolutionary expropriation of all private property in land and resources. If we could start from a blank slate I would be tempted to keep land and minerals as public property. But for many environmental goods, previously free but increasingly scarce, we still do have a blank slate as far as ownership is concerned. We must bring increasingly scarce yet unowned environmental services under the discipline of the price system, because these are truly rival goods whose use by one person imposes opportunity costs on others.[15] But for efficiency it matters only that a price be charged for the resource, not who gets the price. The necessary price or scarcity rent that we collect on newly scarce environmental public goods (e.g., atmospheric absorption capacity, the electromagnetic spectrum) should be used to alleviate poverty and finance the provision of other public goods.

The modern form of the Georgist insight is to tax the resources and services of nature (those scarce things left out of both the production function and GDP accounts) and to use these funds for fighting poverty and for financing public goods. Or we could simply disburse to the general public the earnings from a trust fund created by these rents, as in the Alaska Permanent Fund, which is perhaps the best existing institutionalization of the Georgist principle. Taking away by taxation the value added by individuals from applying their own labor and capital creates resentment. Taxing away value that no one added, scarcity rents on nature's contribution, does not create resentment. In fact, failing to tax away the scarcity rents to nature and letting them accrue as unearned income to favored individuals has long been a primary source of resentment and social conflict.

Charging scarcity rents on the throughput of natural resources and redistributing these rents to public uses can be effected either by ecological tax reform (shifting the tax base away from value added and on to throughput), or by quantitative cap-and-trade systems initiated by a government auction of pollution or depletion quotas. In differing ways each would limit throughput and expansion of the scale of the economy into the ecosystem, and also provide public revenue. I will not discuss their relative merits, having to do with price versus quantity interventions in the market, but rather emphasize the advantage that both have over the currently favored strategy. The currently favored strategy might be called "efficiency first," in distinction to the "frugality first" principle embodied in both of the throughput-limiting mechanisms mentioned above. By "frugality" I mean non-wasteful sufficiency rather than "meager scantiness."

"Efficiency first" sounds good, especially when referred to as "win-win" strategies or more picturesquely as "picking the low-hanging fruit." But the problem of "efficiency first" is with what comes second. An improvement in efficiency by itself is equivalent to having a larger supply of the factor whose efficiency increased. The price of that factor will decline. More uses for the now cheaper factor will be found. We will end up consuming more of the resource than before, albeit more efficiently. Scale continues to grow. This is sometimes called the "Jevons effect." A policy of frugality first, however, induces efficiency as a secondary consequence. Efficiency first does not induce frugality—it makes frugality less necessary—nor does it give rise to a scarcity rent that can be captured and redistributed.

I am afraid I will be told by some of my neoclassical colleagues that frugality is a value-laden concept, especially if you connect it with redistribution of scarcity rents to the poor. Who am I, they will ask, to impose my personal elitist preferences on the democratic marketplace, blah, blah, etc., etc. I am sure everyone has heard that speech. The answer to such sophistry is that ecological sustainability and social justice are fundamental objective values, not subjective individual preferences. There really is a difference, and it is past time for economists to recognize it.

VII. Conclusion

Reducing poverty is indeed the basic goal of development, as the World Bank now commendably proclaims. But it cannot be attained by growth for two reasons. First, because growth in GDP has begun to increase environmental and social costs faster than it increases production benefits. Such uneconomic growth makes us poorer, not richer. Second, because even truly economic growth cannot increase welfare once we are, at the margin, producing goods and services that satisfy mainly relative wants rather than absolute wants. If welfare is mainly a function of relative income, then aggregate growth is self-canceling in its effect on welfare. The obvious solution of restraining uneconomic growth for rich countries, so as to give opportunity for further economic growth, at least temporarily, in poor countries, is ruled out by the ideology of globalization, which can only advocate global growth. We need to promote national and international policies that charge adequately for resource rents, in order to limit the scale of the macroeconomy relative to the ecosystem and to provide a revenue for public purposes. These policies must be grounded in an economic theory that includes physical throughput among its most basic concepts. These efficient national policies need protection from the cost-externalizing, standards-lowering competition that is driving globalization. Protecting efficient national policies is not the same as protecting inefficient national industries.

Notes

An earlier version of this essay was given as an invited address to the World Bank in Washington, D.C., April 2002.

1. It also puts the future at a disadvantage—the present could bequeath an ever smaller throughput and claim that this is sufficient for non-declining utility provided future generations take full advantage of possibilities of substitution in both production and utility functions that preserve natural throughput. But if these substitution possibilities are so easy to foresee, then let the present take advantage of them now, and thereby reduce its utility cost of a given throughput bequest.

2. The throughput is not only measurable in principle but has been measured for several industrial countries in the pioneering physical accounting studies published by World Resources Institute in collaboration with Dutch, German, Japanese, and Austrian research institutes. See *Resource Flows* (Washington, D.C.: WRI, 1997) and *The Weight of Nations* (Washington, D.C.: WRI, 2000).

3. Investing nonrenewable resource rents in renewable substitutes is a good

policy, with impeccable neoclassical roots, for sustaining the throughput over a longer time period.

4. Evidence that growth in the United States since the 1970s has likely been uneconomic is presented in H. Daly and J. Cobb, *For the Common Good* (Boston: Beacon Press, 1989, 1994). See their appendix on the Index of Sustainable Economic Welfare.

5. If welfare is a function of relative income, and growth increases everyone's income proportionally, then no one is better off. If growth increases only some incomes, then the welfare gains of the relatively better off are cancelled by the losses of the relatively worse off.

6. Instead of "deluding themselves" perhaps I should say "being deluded" by IMF and World Bank economists who require this misleading system of national accounts of them. "Hicksian income" is the maximum a community can consume this year without reducing its capacity to produce and consume the same amount next year.

7. Macroeconomists do recognize that the economy can grow too *fast* when it causes inflation, even though the economy can never be too *big* in their view.

8. How might capital flows be restricted? A Tobin tax (a small tax on all foreign currency purchases); a minimum residence time before foreign investment could be repatriated; and most of all something like Keynes' International Clearing Union in which multilateral balance on trade accounts is encouraged by charging interest on both surplus and deficit balances on current account. To the extent that current accounts are balanced, then capital mobility is correspondingly restricted.

9. If your uncle in Texas discovers oil on his ranch, and Texaco is prepared to pay him for the right to extract it, is that not a positive price for petroleum *in situ*? It looks like it, but the amount Texaco will pay your uncle is determined by how easy it is to extract his oil relative to marginal deposits. Thus it is labor and capital saved in extraction that determines the rent to your uncle, not the value of oil *in situ* itself, which is still counted as zero.

10. Nicholas Georgescu-Roegen, *The Entropy Law and the Economic Process* (Cambridge, Mass.: Harvard University Press, 1971).

11. Jeffrey M. Perloff, *Microeconomics*, 2nd ed. (Boston, Mass.: Addison-Wesley, 2000).

12. Some readers may rush to the defense of the textbook and tell me that the production function is only describing value added by L and K and that is why they omitted material inputs. Let me remind such readers that on the previous page the textbook included material inputs, and further that the production function is in units of physical quantities, not values or value added. Even if expressed in aggregate units of "dollar's worth," it remains the case that a "dollar's worth" of something is a physical quantity.

13. I should say that I am thinking of the unit process of production—one laborer with one saw and one hammer converts lumber and nails into one doghouse in one period of time. We could of course multiply the unit process by ten and get ten doghouses made by ten laborers, etc. My point is that the unit process of production, which is what the production function describes, involves no multiplication.

14. Of course one might imagine entirely novel technologies that use totally different resources to provide the same service. This would be a different production function, not substitution of factors within a production function. And if one wants to induce the discovery of new production functions that use the resource base more efficiently, then it would be a good idea to count resources as a factor of production in the first place, and to see to it that adequate prices are charged for their use! Otherwise such new technologies will not be profitable.

15. For example, rents can be collected on atmospheric sink capacity, electromagnetic broadcast spectrum, fisheries, public timber and pasture lands, offshore oil, rights of way, orbits, etc.

5

Placing the Soul

An Agrarian Philosophical Principle

Norman Wirzba

> In abandoning the world we are lost; we are lost again and again. We may speak poignantly of the experience of being lost; but we cannot be clear about ourselves and our situation in so far as our thinking is dominated by that experience. Disillusionment with the world knows nothing of the sacrament of co-existence. It can find no place for the sacramental act. It can conjure out of itself no philosophy of action, for its ultimate implication is inaction.
>
> —Henry Bugbee, *The Inward Morning*

In the opening lines to *The Unsettling of America* Wendell Berry observes that "one of the peculiarities of the white race's presence in America is how little intention has been applied to it. As a people, wherever we have been, we have never really intended to be."[1] Though Berry is quick to note that from a historical point of view this is "too simply put"—after all, there are examples of people who have been and continue to be devoted to the places and communities of which they are a part—nonetheless, there is a longstanding philosophical and religious tradition that has encouraged the training of our attention, care, and desire *away from* this world and this life. In other words, our tendency has at times not only been to forsake a given place for the opportunities afforded by a virgin frontier, but to forsake all physical places as inherently limiting, defective, or beneath our ultimate concern. Moreover, this "otherworldly" thrust has been deeply influential in Western culture. How has this happened, and how is this tradition to be addressed? Is its influence the precursor to our own time's dominant feelings of lostness, boredom, insecurity, and disaffection? How does this otherworldliness affect our ability to care for the earth and for each other?

Philosophically speaking, we can turn to the influential legacy of Socrates and Plato for an enduring expression of otherworldliness. One of Plato's dominant themes is the "care of the soul." Though there are several dimensions to this care, one of the most basic has to do with a distrust of the body, indeed, a distrust of all things material. In the *Phaedo*, for instance, Socrates is reported to say that our bodies get in the way of the pursuit of truth, because bodies invariably confuse and tempt us with pleasures and desires that are fickle, ephemeral, and dangerous. The body fills us with wants, fears, and all sorts of illusions that invariably put us on a path toward disappointment, frustration, and war—war within ourselves over what we crave, and war with others as we compete for the wealth and comfort that comes from the acquisition of material goods.

Clearly this sentiment contains a good amount of wisdom. Our craving for material things and for the pleasures of the body lies behind a great deal of human pain and violence. Consider as just one instance the violence against women (self and other inflicted) that follows from both the glamorous/industrial idealization of what bodies should look like and the reduction of femininity to sexual possession and gratification. Here we can plainly see that a false estimation of bodies can cause tremendous trouble.

Plato, however, goes much further, and in this advance he reflects a predominant Greek philosophical posture. Our problem is not simply the false estimation of bodies and material things: there is something wrong with bodies and things *in themselves*. Thus his counsel that "we must escape from the body and observe matters in themselves with the soul by itself" (66d). The true lover of wisdom "persuades the soul to withdraw from the senses . . . to trust only itself and whatever reality, existing by itself, the soul by itself understands, and not to consider as true whatever it examines by other means, for this is different in different circumstances and is sensible and visible, whereas what the soul itself sees is intelligible and invisible" (83a–b). Failure to extricate the soul from the pleasures, pains, and desires of the body can lead only to suffering and "extreme evil."

The *Phaedo* expresses in direct, simple form an idea more completely expressed later in the mature work of Plato's *Republic*, one of the most influential of the Platonic dialogues. Here Plato constructs a metaphysical picture of reality in which otherworldliness is central: what is ultimately real and truly good cannot be understood in terms of our natural lives or through ordinary human experience, since this world is characterized by mutability and flux, instability and contingency, limits and partiality, i.e., imperfection. Owing to the inherent dissatisfaction that must inevitably follow from our experience with things in this life, the true philosopher must instead seek eternal goodness, beauty, and truth in another realm, a realm that transcends this world and its faults. As expressed in the analogy of the "divided line," reality can be broadly split into two parts: one referring to the sensory realm, the other to the supersensory or intelligible realm. The goal of philosophical dialectic is to take us beyond the sensory so we can approach the eternal "forms" or "ideas" of things in their unchanging, universal essence. In order to access (with our souls) the world of being itself, which has as its summit the Good, we must leave behind the world of becoming (518c).

When this mature metaphysical account of reality was combined with Greek Orphic religion espousing a rigid soul/body distinction, it became clear that the goal of all human life must be to downplay, literally degrade, bodies and material things, for they are at their best merely the temporary prisons of a divine soul, and at their worst they are the obstinate impediments to the soul's salvation, the soul's release from the body so that it can enjoy eternal life with the gods. On this Greek account, human life is good and praiseworthy insofar as it signifies a carefully measured and reasoned disdain for a fluctuating and impermanent world.

Importantly, this is not the whole story to Platonic or Greek philosophy. Indeed, Plato and Aristotle reserved an important role for the practical arts in understanding and developing what is good and excellent in this life, and Plotinus, later on, would hold suspect any attempt to confine the spiritual to a supraterrestrial place. Nor should this account be understood to suggest that there is nothing worth salvaging in Platonic or Greek thought. Rather, my point has been to show how one of the most influential strands of "otherworldly" thinking grew out of legitimate and serious philosophical exploration into the nature of a good life.

Modern thought, though it may have prided itself on abandoning the mythological frameworks of the ancients, continued to promote the deprecation of the material world, not by enjoining the soul to flee this life, but by denuding materiality of all inherent value. Eliminating natural teleology, the idea that natural things have an end or purpose internal to themselves, made possible the belief that human minds are the sole carriers of value, the origin and end of all purpose, and thus are mandated to do with bodies of all kinds whatever they deem useful or pleasing. Because the natural or physical world is without intrinsic value (it is brute matter to be evaluated and handled in quantitative rather than qualitative terms), it cannot possibly serve as a source for moral deliberation. From this presupposition the "is"/"ought" distinction that underlies so much modern moral theory—the idea that the moral "ought" cannot be derived from the way the world in fact "is"—easily follows. The natural world cannot guide moral striving because it is without moral charge. Given this split between amoral facts and mind-bestowing value, a split mandated by the scientific method, we should not be surprised that people are assumed to be disembodied minds. We cease to be biological organisms intimately tied to (and thus limited by) supporting habitats, but instead become machines inhabited by ghosts. Freed of all material/natural constraints, we are now in the position to "rise above" the world and remake it and ourselves according to our own liking.

In postmodernity the quest to free humanity from all natural and normative constraints, to free us of the concerns of place, becomes even more striking. Many today would acknowledge that we live in a time dominated by infinite possibilities *and* uncertainty about ultimate ends. Increasingly we live as individuals (all the while sensing the risk and anxiety that accompany our attempt to make ourselves the ultimate court of appeal), because there is no public sphere in which ends and values can be defended as universally binding. In the advanced stages of global capitalism we are now all ultimately shoppers, individually choosing products, values, and our own identities. Life is reduced to scanning for possibilities, seeking the ever new and improved as we superficially evaluate our choices largely in accordance with the changing fashions of media-manufactured desire. We operate above or apart from the world, within the sphere of our desires and wishes, entering into reality when it is time to fill our shopping baskets.[2]

In important respects, our contemporary situation reflects a continuation of Socratic otherworldliness, now expressed in a different key. Though we may have given up the Greek idea of the soul's union with eternal truth and blessedness, we have not abandoned the flight mechanisms that would enable us to transcend this world. We have not embraced materiality or the demands of place. As postmoderns we too disdain the binding pull of bodies and the natural world. We engage reality on our own terms rather than the terms set by physiology, biology, ecology, meteorology, and communal tradition.[3]

Western religion, particularly various expressions of Christianity, has also promoted forms of otherworldliness. In part this stems from the pervasive influence of Platonic and neo-Platonic thought on the Christian traditions. But it also stems from an ambiguity and tension sitting at the heart of the Christian faith. On the one hand (here borrowing heavily on the Hebrew faith out of which Christianity grew), God has created a world that is good and to be cherished as an expression of worship and gratitude to God. But on the other hand, this world is the site of so much sin and wrong-doing that it needs to be transformed (perhaps destroyed) and rebuilt so that it might better reflect God's intention and rule. Thus, the Christian does not simply live in one world (or "city" as Augustine famously put it), but rather two, or perhaps more properly, *between* two. Christians are forever "on the way," on a journey to their proper abode as it is revealed to them by God in the life of Jesus Christ and the teachings of the church. How this "way" was initially conceived came to have tremendous implications for the Christian understanding of a believer's place in this life.

Some Christians, particularly those influenced by apocalyptic modes of thought that stressed the imminent return of their Lord, but also those influenced by Gnostic tendencies that maintained a strong soul/body dichotomy or Docetic believers who claimed that Jesus Christ only "seemed" to take on human form and suffering, came to believe that this world was ultimately of little value. What mattered most was their eternal home with God. Given that many Christians experienced their daily lives as filled with suffering and toil, we should not be surprised that we see down through the ages ample evidence of the desire to flee this life so that union with the divine might occur.

Numerous texts, ranging from the early church to the contemporary fascination with "end times" literature, could be gathered to show forth this tendency. But so too could many of the church's practices, as when medieval clerics and monks played important roles developing the technologies and the mentality that would foster an instrumental approach to the natural world (nature exists to serve human interests and needs), and when millennial hopes are employed to justify the using up or the destruction of natural habitats so that God's reign might begin more quickly. Throughout these texts and practices we can see a propensity to view this world and this life as temporary at best, as a burden or obstacle at worst. No wonder, then, that "This World Is Not My Home" is a favorite song for contemporary dejected and oppressed Christians:

> This world is not my home, I'm just a-passing through
> My treasures are laid up somewhere beyond the blue
> The angels beckon me from heaven's open door
> And I can't feel at home in this world anymore.[4]

It is important to understand that this otherworldly stance was not inevitable. In fact, a strong case can be made to show that it violates what is essential about the Christian faith and life. One of the most detrimental effects following from otherworldly Christianity of this sort is an understanding of God as finally detached and unmoved by the affairs of the created order. If this life is little more than a "veil of tears," a place of sorrow and suffering, and if this suffering is inherently bad and to be avoided at all costs, then we should not be surprised to see a fairly disincarnate form of Christianity emerge. A "disincarnate" God, in good Gnostic or Docetic fashion, does not fully or compassionately enter into the world so as to be identified with and redeem it, but rather encounters the world so as to judge it, condemn it, and finally destroy it in a conflagrant spectacle (again, it seems, that materiality and corporeality are themselves a problem). This God, much like the gods of Greek philosophy, and in stark contrast to the incarnate God who chose to "dwell among us" and share in our fate (see John 1:14 and Revelation 21:3), is finally a God of self-containment, a God who is indifferent to and has no need of the creation. God's perfection simply cannot abide such manifest imperfection. The Christian's best hope, in this context, is to be one of the fortunate few who escape this earth before the conflagration begins.

Can we clearly perceive ourselves and our situation so long as various forms of otherworldliness dominate our thinking? Or to put the point more practically, can we *properly* engage the world if we despise the bodies in terms of which such engagement occurs, or despise the natural bodies upon which our own lives so clearly depend? One of the lasting contributions of *The Unsettling of America* was to show that on both counts the answer is a resounding *No!* Though we might dream of ourselves as disembodied, immortal souls, or as complex computers that will finally shed all biological and physiological limitations, the fact remains that we live necessarily through our bodies. And these bodies, in turn, necessarily live through the bodies of others—wheat, rice, steer, fish, microorganisms, bees, chickens. We simply cannot avoid or override the ecological truth that "our land passes in and out of our bodies just as our bodies pass in and out of our land," and that all the living "are part of one another, and so cannot possibly flourish alone."[5] Insofar as an otherworldly mind-set diminishes or denies human embodiedness and embeddedness, our ability to live honestly and responsibly in the here and now is severely compromised.

If we are to live properly and responsibly, with an honest appreciation of who we are and what our situation recommends, we must start by placing our souls firmly within the contexts of bodies and things, just as we should place culture securely within the land (*ager*). We must halt the flight instinct that would free us of this world. Such placement, rather than putting an end to the sense of ourselves as spiritual beings, recasts philosophical and religious life in a more appropriate light. Placing the soul in the plenitude of materiality, in other words, has the potential to rescue the mind from the disorder and confusion that invariably follow its severance from its indispensable roots and limiting conditions.[6] Not only will such placement return spiritual reflection to what is perhaps its most ancient conviction—"To be is to be in place"[7]—it will also give to philosophical and religious quests the integrity they often lack.

Since the whole of philosophy cannot be addressed here, what remains is for us to consider how the contour or shape of philosophical reflection is transformed in the wake of the soul's placement in the material or natural world. If we take as our starting point the standard definition of philosophy as the love (*phileo*) of wisdom (*sophia*), we do not get very far unless we consider what love and wisdom first entail. What sort of love and wisdom are we talking about, and how does this love affect the way wisdom emerges? If we turn to ancient models of philosophical reflection, what becomes clear is that the philosopher was first and foremost interested in practicing a way of life.[8] To be sure, philosophers could develop highly sophisticated theoretical accounts of the nature of reality and of our ability to cognitively grasp the world. But this cognitive work was secondary and in the service of forming good citizens and good souls, of elucidating an ideal human life as it should be lived in the here and now (Plato's immortality of the soul notwithstanding), and in the light of eternity. In other words, philosophical reflection was intimately tied to experience, to the testing, trying, and experimenting of life that constitute our condition.[9]

While this may seem obvious, it is striking to see how many philosophers, particularly in the modern period (the era in which the scientific method fully develops), divorced reflection from concrete experience. In the desire to be "objective" or "detached" observers of the world, assuming that only through such detachment can the truth of things be known, what is forgotten is that our thinking is never merely "about" the world, but also "from" the world. As living beings we cannot simply be spectators of a world, as if we did not breathe its air, drink its water, ingest its material nutrients, or consume its goods. Our lives are literally and figuratively drawn from the earth through our bodies. At the very least, this means that the quest for philosophical truth demands an open life (since it is dependent on a porous body)—not simply an open mind—sensitive to the many ways in which our bodies and our minds are nourished and supported by the lives of others. To close ourselves off from the manifold richness and diversity of our spiritual and physical interdependencies, or to be ignorant or disdainful of the limits and possibilities of place, would signal philosophical failure. To be a genuine philosopher requires that we become more attentive to our bodies and attachments, and then commit and abandon ourselves to the experiences of life—the flows of birth, growth, disease, and death—realizing that the experiences themselves are not our possessions or easily within our control. Lest we dismiss such abandonment as overly mystical or sentimental, we should understand it in terms of the sympathy of love, the training of our desire on the need and well-being of another. We care for people, billions of organisms, and the myriads of habitats they support, because we now appreciate that we all draw our life from each other, and that we are all mutually implicated in each other's fate. Abandonment thus includes enchantment ("falling in love" with the splendor and beneficence of a creation finely made), but also tenacity and commitment (the "work" of love that overcomes ego-centeredness so that the needs of others can be addressed).

If concrete, embodied experience is to be our point of departure, then we must be careful to note the conditions necessary for the honest and responsible perception, reception, and engagement with reality. Clearly, something like a furtive glance will hardly suffice, since there is little opportunity here for sustained or detailed attention. Similarly, a perception of the world that is geared toward its utilitarian or instrumental use will also fall short, because in this case things are not experienced in their intrinsic uniqueness or individuality, but in terms of their usefulness to us. As Michel de Montaigne put it (and as Berry begins *The Unsettling of America*), "Who so hath his mind on taking, hath it no more on what he hath taken."

Insofar as we desire to understand the world (Aristotle claimed that this desire is a universal one and the original source of philosophical reflection), we must be prepared to commit ourselves to the world in all its incomparable uniqueness and particularity. Bugbee argues that "the measure of our *understanding* of reality lies in our capacity for the responsible realization of unqualified affirmation."[10] Unqualified affirmation presupposes that we allow the world of things and bodies to become fully present to us, to let their demands be felt by us, and that we not distort or block—whether through fear, laziness, or arrogance—the stream of reality that continually surrounds us. We must also learn to respect the integrity of things by giving them the space to be themselves. In this hospitable gesture we show our commitment to be considerate and sympathetic *Homo sapiens* whose knowledge grows out of sustained and intimate contact with the wider world (remembering that the term "sapiential" refers to the intimate knowledge of "tasting the other").

The foregoing remarks on the nature of the philosophical posture have suggested that we cannot embark on the quest for truth if we do not at the same time commit ourselves to the full disclosure of reality as it confronts us. What we must now realize is that such a posture is not possible without love, for it is in terms of love that the true marks of knowing can emerge: openness, affection, resilience, patience, humility, vulnerability, kindness, intimacy, responsibility, and perhaps most important, repentance. In other words, lovers of wisdom must first be lovers in the most genuine sense of the term, people who are considerate and show compassion for the one loved.

The religious and poetic mind has most often made the clear identification between wisdom and love. Spiritual classics abound with the admonition to seek experience rather than accumulated knowledge and to follow love in the fields of experience: knowledge puffs up the knower, but love builds up what is known. Knowledge per se isn't evil; rather, love, because it is attuned to the goodness and graced character of things, is in a better position to appreciate their truth.

We see this especially well in the fourteenth-century Franciscan writer Duns Scotus, the "subtle doctor." For Scotus it is God's relation to the creation—seen most clearly and in its most perfect form in the incarnation of God in Jesus Christ—that gives to things their purpose or goal.

But this is not a deistic God who, sometime in the past, operated on the world from outside. Rather, God is at work in all things at all times, sustaining and enabling them to be the things that they are. Indeed, given Scotus's high affirmation of things in their particularity, what he called their *haecceitas*, each and every thing serves as the vehicle for the revelation of God. All things share in the divine life, and all things express the divine life. That anything exists at all testifies to a divine, creative love.

Our access to this world of particulars, however, cannot be true and responsible if in our knowing we subvert the ability of things to be themselves so that they might instead be what we want them to be. Things are important in themselves. They have value—"finality," as Bugbee would put it—independent of whatever value we may assign to them. Thus, if we want to know things we must first love them by attending to them and relinquishing our fears, desires, and self-serving agendas. For Scotus, the life of Christ serves as the model for this loving approach to things. Christ is the true philosopher because he embodies in his ministry the welcoming and caring reception of others so that they might more fully be the beings they are meant to be. Indeed, in the Christlike effort to understand, serve, heal, feed, and reconcile the earth and its communities we show forth the highest wisdom.

To speak of the priority of love is to raise anew the question of how we should act, for the genuineness of love requires its concrete expression in action. Wisdom is the reflection of this love in action. Bugbee writes, "The understanding of reality and human fulfillment are bound up with effective action. . . . Thus, apart from one's own commitment, one's own complete commitment in action, the distinct things which we know, however systematically conceived, must remain as noise; they fail to make ultimate sense."[11] The authenticity of our understanding, in other words, is inextricably tied to the authenticity of our lives, to what Berry calls the propriety of our lives. Propriety is concerned with how well we "fit" into the contexts we find ourselves in; consequently, propriety becomes a question of whether or not our actions respect and maintain the conditions and connections that make our lives possible. "To raise the issue of propriety is to deny that any individual's wish is the ultimate measure of the world."[12] Proper understanding is thus always attuned and responsible to the variety and demands of the habitats in which we move. This is why moral reflection, one of the primary activities of the soul, must also be firmly placed in the world. As Berry notes, "How you act *should* be determined, and the consequences of your acts *are* determined, by where you are. To know where you are (and whether or not that is where you should be) is at least as important as to know what you are doing, because in the moral (the ecological) sense you cannot know *what* until you have learned *where*. Not knowing where you are, you can make mistakes of the utmost seriousness: you can lose your soul or your soil, your life or your way home."[13]

The modern image of the knower and the ethical agent as one who constructs a blueprint of how the world ought to be, and then remakes the world according to his or her own design, simply misunderstands and forfeits a philosophical life. This is life without wisdom, guilty of a basic impropriety that confuses wisdom with the possession of a body of knowledge. Wisdom, more properly conceived, has to do with the patience, courage, and strength we need to remain true to our situation and condition as we work our way through it. Put in most simple terms, wisdom is the capacity to remain faithful and true to reality as we encounter it, without falsifying, evading, or destroying it. The wise one is the one who manifests in his or her life full responsibility for the world as it has been given and received.

As has already been suggested, one of the hallmarks of the modern and postmodern worlds that makes acquisition of such wisdom especially difficult is our growing disillusionment and disenchantment with the world. Fueled by otherworldly attachments or by anxiety, boredom, and disaffection in the face of a valueless universe, we find fewer instances in which people deeply or responsibly love their bodies, their homes, or their habitats. We see this in the growing trend of self-mutilation practices among young adults and in the ashamed desire to make our bodies something different (hopefully more sexually appealing) than what they naturally are. We see it also in the inability of most people to care for and maintain their living spaces (for these tasks we hire professional cleaners, fixers, builders, and sanitation experts) and their communities (for these we hire childcare workers, therapists, and nursing home providers). What is lacking is the sense of the abiding connection between ourselves and our worlds; because of this disconnection we cannot exercise the virtues of love (such as attention, patience, affection, resilience) that would enable us to be the caretakers of the world and ourselves that we should be.

Our love, in other words, has become abstract, cut off from a deep (and practical) immersion in and commitment to place and community. What we fail to see, oftentimes, is how a ubiquitous consumer mentality directly contributes to this abstraction. Consumerism does not refer simply to the purchasing of many (often unnecessary) things. It is, rather, an approach to reality that fundamentally alters the ways we engage and relate to the world around us. As consumers our attention is focused primarily on obtaining or anticipating (since the future is the primary temporal mode) for ourselves the commodities that will satisfy desires manufactured and induced by the market. Our engagement with an external world, now increasingly characterized in terms of commodity exchange, has less to with reality itself than with marketable images that determine production and spending. Our perception and reception of the world are thus made oblique. We do not encounter reality on its terms, but in terms of the much narrower orbit of market concerns. A consumer mentality, in other words, contributes to our overall ignorance about the truth of reality, just as it works against a life of wisdom, because we now relate to the world more ephemerally as the scanners and purchasers of it.

This way of speaking will sound strange only until we consider the fact that in our time, when we claim to "know" more about the world than ever before, we are also responsible for its most widespread and systemic exhaustion and destruction. Our much trumpeted knowledge has not led to a sympathetic or affectionate understanding that would induce us to take care of the world, because our knowledge is the sort that is content with knowing *about* rather than knowing from within, from the perspectives of intimacy and practical engagement, the character of the world. Our knowledge has become increasingly economic and simplistic, reducing things to their exchange value, and thus abstract, much like our love. It has become improper and without art, oblivious to whether or not our understanding results in a greater "fittedness" with the wider world.

We can see how a consumer society and mentality contributes to abstraction if we compare it to the disciplines of production as they have been practiced in many traditional and indigenous cultures.[14] To be a producer, an artisan for example, is to submit oneself to a socially/culturally defined discipline or craft that requires extensive training and patience. One must learn the art of design, which means that one must gain a sense for the deep reality of things, have a qualitative grasp of things as they naturally are.[15] To know the nature of things means that one grasps their essential and manifold characteristics as they connect with the world of which they are a part. Building on Aristotle's famous account of how knowledge is achieved, we can say that someone who genuinely knows will be able to answer the following questions: What type of material will make the product good? What are a material's (and habitat's and community's) limiting conditions? What methods of production will yield the best (and the most sustainable and safe) results? To what end or purpose should products be made, and how well does this goal fit with broader social and ecological ends? What form or design will best promote quality, durability, beauty, functionality, i.e., excellence? To answer these questions is to enter practically into a complex, moral dimension that takes seriously the places we are in and the character of our dwelling within these places. They demand the sort of democratic and public conversation that has all but disappeared in our time. None of these questions can be adequately or truthfully answered without patient, detailed attention to place, or without sustained commitment to place and community. True knowledge and understanding grow out of our productive engagement with the world, the engagement serving as the corrective and guide to our fanciful or flight-prone ways.

The contrast between a producer and consumer is thus fundamental. Insofar as we are primarily consumers of the world (engaging reality in terms of our desires and wishes), we limit and distort our knowledge of it, and thus our ability to care for it properly. But as we take up productive roles, become active participants in the construction and maintenance of the flows of life—as when we grow food, become intentional about parenting, celebrate communal contributions, and develop a sense of civic responsibility—the claims and benefits of place will become more richly felt and appreciated. This is not to say that we will cease to be consumers, or that we will all suddenly become good. Rather, we will become responsible consumers who now more honestly appreciate the costs and requirements associated with living.

The difference becomes clear if we consider the example of food. To grow food requires that one become knowledgeable about biological and chemical processes, that one be attentive to topography and weather, that one be mindful of the particularities and peculiarities of place, plant, and community. Success is thus directly tied to our ability to get our egos out of the way and fit in and work with natural processes going on around us. To eat the food one has grown is thus to become aware of the gifts and limits of place—we cannot master growth, only gratefully assist and receive it as it comes—and the costliness of those gifts, since the processes of life are always intertwined with the processes of death. If we are merely the consumers of food, we will fail to appreciate these costs, and thus more likely take for granted or abuse the natural contexts that make food production possible. But, actively involved in food production, we will come to see our own lives as enveloped in a much larger drama that is life-giving but also vulnerable to exhaustion and destruction. The responsible, sacramental sense that we must care for this natural drama, see that it is maintained and not destroyed or compromised, will be a natural outgrowth of our sustained engagement and work with it.

What this means for the moral life is that the matter of first, and perhaps greatest, importance is that we not think of ethics as primarily constructing blueprints for action, particularly if these blueprints are drawn up by a disembodied or disenchanted mind. Rather, the most important task will be the "immediate clarification with regard to a foundation of life that is absolutely genuine (as opposed to optional, arbitrary, or conditional), and utterly beyond artifice or manipulation." What Bugbee means is a life that is devoted to the limits and possibilities of particular places and that is prepared to learn from those places what is proper and improper. Leading such a life is by no means easy, for our temptation is invariably to impose with undue thought or consideration our own ways. Hence Bugbee continues: "Except as ethical reflection is undertaken in what I must call the spirit of prayer—an utmost form of commitment which cannot be simulated or induced—it cannot be freed from arbitrariness."[16] Utmost commitment stands in stark contrast to the ephemerality of a shopping mentality that is prepared to forsake the old for the new with the slightest marketing provocation.

To speak of morality in this way is to highlight its provisional character, its vulnerability and fragility. To be moral is to live with the sense of how often we have erred in our presumption to know how we and the world should be (the carnage of human history and the destruction or exhaustion of untold habitats should be enough evidence for that). It is to accept responsibility for the harmful effects of what we have done. The Good, not being something we can understand (and control) in abstraction, must grow out of a sense of graciousness that dawns on us in the midst of the trials and turns of action, which is to say through human experience. It is only as we are faithful to the particularities and demands of place and accept responsibility for our actions in those places, that we can claim to be moral beings at all.

At various points I have referred to the honest response to the world as a faithful response. To speak this way is, of course, to reintroduce a religious category into the heart of the philosophical enterprise. This faith, however, is not the simple adherence to dogma, as if faith were finally or maximally a matter of intellectual assent to a series of religious propositions. We do a better job making sense of this faith if we relate it to the forms of love we have described as necessary ingredients on the way to wisdom. To be faithful to the world is to trust and accept responsibility for the gifted and graced character of experience, to immerse ourselves into the flow of experience and there find ourselves maintained by meaning and love that we do not control and, for the most part, do not deserve. To speak of meaning and love in this way is not to deny the often tragic character of our existence; rather, it is to recognize in tragedy the ambiguity and pain that accompany all human effort to live appropriately in a world of limits and possibilities.

To be faithful to the world does not commit us to idolatry or paganism. Idolatry would represent a mentality in which the world of things, or a particular thing, becomes the object of our devotion for purposes of human salvation or consolation. An idol, in this case, completes what we find lacking in ourselves by isolating and magnifying an element of the world to suit a human end. The more authentic encounter with the world will result not in idolatry but in the sense of the sanctity of things. Sustained, patient, responsible engagement with things invariably leads to the sense that at root the world we know remains unknown, beyond our comprehensive or utilitarian grasp. Hence Bugbee maintains that even our everyday experiences, like washing clothes or preparing a meal, when discerned in their appropriate depth, point us to wilderness. Faith and reflection begin with the experience of wonder, with the experience of Job in the Hebrew scriptures, who found in the bedazzlement of creation's variety and care the proper purification and focus for his own thought and action.

There is, as Gerard Manley Hopkins observed in the poem "God's Grandeur," "the dearest freshness deep down things." To the religious mind this freshness corresponds to the vivifying presence of the divine creative power. To the philosophical mind it refers to the newness of possibility that remains deep within the world. To sense this freshness, to see in ourselves and in our neighbors and neighborhoods this possibility, is to find existence a miracle and an adventure. Before this miracle we cannot remain passive or bored. Before it we must give thanks and commit ourselves to the responsible work of caretaking and celebration. Only then will the soul, now properly situated in body and place, find its true peace.

NOTES

1. Wendell Berry, *The Unsettling of America: Culture and Agriculture,* 3rd ed. (San Francisco: Sierra Club Books, 1996), 3.

2. For a sociological description of how various forms of modernity affect personal and social self-understanding see Zygmunt Bauman's *Liquid Modernity* (Cambridge, England: Polity, 2000).

3. There is a recent tendency in philosophy and sociobiology to say humans are nothing more than complex physiology interacting with a material and chemical world. This reductive move should be resisted since it bars the possibility of a strong moral evaluation of human agency (there is nothing about the materiality

of the world itself that recommends our caring for it). The soul, understood as the center of responsible agency, thus needs to be "rescued" and "placed" at the same time.

4. J.R. Baxter Jr., "This World Is Not My Home" (Dallas, Tex.: Stamps-Baxter Music and Printing Co., 1946).

5. *The Unsettling of America*, 22.

6. Remembering here the Augustinian maxim: "The punishment of every disordered mind is its own disorder" (*Confessions*, 1.19).

7. This point is worked out in great detail in Edward Casey's *Getting Back into Place: Toward a Renewed Understanding of the Place-World* (Bloomington: Indiana University Press, 1993), 14, and in its companion historical work *The Fate of Place: A Philosophical History* (Berkeley: University of California Press, 1997).

8. The definitive account for this view of ancient philosophy is Pierre Hadot's *What Is Ancient Philosophy?* (Cambridge: Harvard University Press, 2002).

9. In my account of philosophical reflection that is attuned to experience I will draw frequently on the careful, nuanced study made by Bugbee in *The Inward Morning: A Philosophical Exploration in Journal Form* (1958; reprint, Athens: University of Georgia Press, 1999), where he writes: "Experience is our undergoing, our involvement in the world, our lending or withholding of ourselves, keyed to our responsiveness, our sensibility, our alertness or our deadness" (41).

10. Ibid., 75.

11. Ibid., 85.

12. Wendell Berry, *Life Is a Miracle: An Essay against Modern Superstition* (Washington, D.C.: Counterpoint, 2000), 14.

13. Wendell Berry, "Poetry and Place," in *Standing by Words* (San Francisco: North Point Press, 1983), 117.

14. My intent, clearly, is not to disparage consumption per se, since we all need to ingest the gifts of the earth to live. Rather, my point is that a consumer mentality, as it has developed recently, fundamentally alters the character of consumption and production however and whenever they occur.

15. Consider here Ananda Coomaraswamy's succinct formulation of a traditional understanding of art: "Art is an imitation of the nature of things, not of their appearances" (*Christian and Oriental Philosophy of Art* [New York: Dover Publications, 1956], 19).

16. *The Inward Morning*, 70.

Part 2

Assessing Our Situation

6

THE CURRENT STATE OF AGRICULTURE

Does It Have a Future?

Frederick Kirschenmann

> If the [current] pattern holds, farming as a way of life will mainly disappear within the next 50 years, large swaths of the country will be virtually depopulated.
>
> Jedidiah Purdy

When in the mid-1970s Wendell Berry was writing his singular work *The Unsettling of America*, the industrialization of agriculture was already well underway. The transformation of agriculture into an industry was enthusiastically endorsed by many agricultural pundits and "experts." In fact, as Berry tells us in the preface to the first edition of *Unsettling*, he was "incited" to begin taking the first notes for his book in 1967 when President Lyndon Johnson's special commission on federal food and fiber policies made its report. In the view of the commission, a major problem with U.S. agriculture was that we still had too many farmers on the land. The "technological advances" had so reduced the need for farm "manpower" that national farm income simply could no longer support as many farmers.

The industrialization of agriculture, particularly after the Second World War, precipitated a free fall decline in farming populations. Indeed, as a study by Calvin Beale has indicated, the decline in farm populations was a trauma not tied to market and price fluctuations, but rather was endemic to the industrialization process itself.[1] The result: "the total number of farms has declined from 6.5 million in 1935 to 2.05 million in 1997, and most of this huge decline took place among family-type farms."[2] The decline in farm population can no longer be seen as an aberration, or a correction to an otherwise healthy system. It is also instructive to recognize that while farmers have been reluctant to adopt industrialization, in the end they have almost always complied. Frederick Buttel points out that farmers have been aware of the spending treadmill that industrialization inevitably puts them on, but economic forces ultimately force them into compliance.[3]

Toward a Bifurcated Food and Farming System

For the most part the situation has continued to deteriorate since the publication of *The Unsettling of America*. Not only have farm numbers continued to decline (with the exception of the very small farms), but we are now faced with major structural changes that threaten to dramatically alter the landscape of rural America. Here is what we are seeing.

We are moving rapidly into a bifurcated food and farming system. At one end of the scale are a decreasing number of increasingly large farms that produce a single, undifferentiated bulk commodity, for a consolidated firm, most often under a contract written to accommodate the business interests of the firm. According to the most recent (1997) USDA statistics, 61 percent of our total national agricultural product is now being produced by just 163,000 farms, and 63 percent of that production is tied to a market or input firm by means of a contractual relationship.

Direct-market farms occupy the other end of the spectrum. These farmers sell their products directly to food customers through various marketing arrangements—farmers' markets, community supported agriculture arrangements (CSAs), direct sales off the farm, home deliveries, and various Internet networks that directly link producers and consumers. This is a rapidly growing sector of the food industry, but it remains a tiny portion of the food and agriculture system. No reliable statistics exist to establish what portion of total agriculture production is served by direct marketing. But since 1.3 million of American farmers, those classified as part-time or retirement or residential farms, account for only 9 percent of the total national agricultural product,[4] we can imagine that the direct-market farmers capture only a tiny percent of that meager margin.

Yet these direct-market farms have increased significantly. CSAs have grown dramatically all across the country and farmers' markets have grown substantially. In California alone, farmers' markets are now reportedly a $100 million business. The number of these farms is likely to continue to grow, especially on the urban fringe and in regions unsuitable for industrial farms. Yet, they will likely always comprise only a tiny segment of production agriculture.

In between these two farm sectors we have approximately 575,000 farms, classified as small to midsize family farms, that produce 30 percent of our total national production. Twenty-seven percent of these farms are tied to a marketing or input firm by means of a contract that determines at least some of the management decisions on the farm.[5] So, while we still have nearly 2 million farmers in America—slightly less than the total number of prisoners housed in our nation's prisons—the majority of our production comes from a handful of very large farms.

These statistics might not be so troubling were it not for their implications for the future. As these trends play themselves out, we are likely to see some dramatic changes on America's landscape that I suspect will be equally disturbing to urban and rural citizens.

Changes on the Landscape

First, the food and agriculture system now emerging from this industrialization process will dictate how farm management decisions are made. Farmers will no longer be able to make management decisions to serve the interests of their farm or the community in which the farm exists. They must make choices that serve the business interests of the consolidated firms with whom they contract. Rapid consolidation, initially in the seed and manufacturing sectors, but now in the food retailing sector, means that about six multinational retail firms will determine not only the size of America's farms, but also the type of management decisions made on those farms.[6]

Independent farmers, selling their production into the free market, have always made on-farm decisions based on a variety of intended outcomes. In addition to managing the farm for profitability, most farmers also made decisions that assured the survival of the farm in its particular community so that it could be passed on to future generations in good health. This is *not* to suggest that small, independent farms have always been managed to prevent soil erosion, protect water quality, or maintain vibrant communities. There is a long history of degradation and loss that belies such a romantic picture of the yeoman farmer of the past.[7] But it *is* fair to say that many independent farmers have included these nobler considerations in their management decisions as a way of insuring the health of the farm for future generations. As farmers now are increasingly forced into contractual arrangements with highly consolidated firms—"forced" because no other markets will remain open to them—on-farm decisions will, of necessity, be made based on the business interests of the firm issuing the contracts. The nobler considerations of the past will be ignored.

The business interests that drive these large consolidated firms are based on three primary objectives: the development of supply or value chains, biological manufacturing, and the reduction of transaction costs.[8] Each of these business objectives will have a profound effect on how local farms are managed.

The *development of supply chains* means that on-farm decisions will no longer be made to benefit the long-term economic viability of the farm, or the quality of life in the community, or the health of the natural resources that sustain the farm. Decisions throughout the supply chain will be made solely to compete effectively with other supply chains and to gain a larger share of the consumer's food dollars.

The introduction of the concept of *biological manufacturing* means that farmers can no longer manage their farms to encourage the normal functions of the animals on the farm, or to enhance the diversity of the landscape, or to promote the general health of the farm. Rather, farm management will necessarily focus on technologies designed to produce uniform products that meet the desired processing and retail objectives of the firm.

And the *reduction of transaction costs* means that consolidated firms will do business only with the largest farmers. It is simply less costly to contract with one farmer who raises ten thousand hogs than to issue contracts to ten farmers who each raise one thousand hogs. All but the very largest farms will become "residual suppliers."

Social and Economic Transformations

Given that farms will be pushed to these new levels of specialization, concentration, and uniformity, profound changes will take place on the landscape. First, farms will be replaced by industrial centers. In Iowa, for example, it is now being suggested that farms of the future will consist of 225,000-acre industrial complexes. Some argue that it will be necessary to consolidate farms into such industrial behemoths to gain access to markets and to negotiate effective prices with input suppliers. This transformation would reduce the number of "farms" in Iowa to 140. Surely such "farms" will not buy equipment from local dealers or fertilizer from local suppliers, nor will they deliver grain to local elevators. As with other industrial complexes, labor will consist largely of minimum-wage earners.

For the most part, commodities produced on these megafarms will be owned by the consolidated firms that issue the contracts. Just as Tyson retains the ownership of chickens placed on "farms" to be raised for them, using *their* feed, managed in accordance with *their* management plan, so other livestock species and patented seed crops will increasingly be owned by the firm, raised for the firm, in accordance with the firm's management plans, using the firm's technology and inputs. In 1992 *Time* magazine had already begun to refer to the "farmers" raising Tyson's chickens for them as "serfs on their own land."[9]

This is a future, in other words, where all local business transactions will be made with distant supply chains, with the benefits accruing to shareholders who most likely live in distant places. And the "farmers" who provide the labor for these operations will be allowed only minimal independent judgment and creativity. Like any other franchised business, such franchise farms will be given the "freedom" to operate in accordance with the firm's directions and to accept most of the liability. In effect, this further transformation from agrarian to industrial agriculture will amount to emptying the landscape of all of its local agriculture-related economies and talent.

BIOPHYSICAL TRANSFORMATIONS

In addition to such social and economic transformations there will be commensurate biophysical transformations on the landscape. We know from past experience that large industrial complexes, owned by absentee landlords and managed by a highly centralized managerial class, do not exhibit a commitment to the care of the environment in which they exist. Witness Love Canal, Louisiana's "cancer alley," the burning Cuyahoga River, the PCBs in the Hudson River, the dried-up Rio Grande River, and the hazardous waste inserted into farm fertilizer in Quincy, Washington. There is no good reason to believe that industrial farm complexes will operate with any higher degree of environmental care than any other industrial complex—indeed factory-farm poultry and hog complexes already serve as harbingers of a clouded future.

Since each of these agricultural industrial complexes will specialize in the production of only one or two commodities, they will foster additional biophysical degradation. We now know that imposing specialization on any ecosystem causes a host of ecological problems. These problems include the elimination of the biodiversity that is essential to the resilience and productivity of any ecosystem.[10] Furthermore, the uniformity and specialization demanded by this new level of industrialization invites genetic uniformity that in turn leads to additional vulnerability. Again the poultry industry presages the future. William Heffernan reports, for example, that "90 percent of all commercially produced turkeys in the world come from three breeding flocks."[11] Such genetic uniformity, initiated to obtain a uniform product, results in birds with such compromised immune systems that their health cannot be maintained without extensive use of antibiotics.

Farms, of course, are ultimately micro-ecosystems that exist within macro-ecosystems. As such, agriculture is an inevitable part of the larger dance of life—part of that complex, interdependent web of life that has evolved (and continues to evolve) over four billion years. We ignore that evolving complexity only at our peril.

The standard industrial answer to this cautionary tale is, of course, that we will always have the technological capability to restore any damage we may do to the ecosystem through our industrialized agriculture—especially with the newly discovered technological capacity of genetic engineering. We now seem to have convinced ourselves that we can redesign life to live better in a new biological order of our own making.

In *The Future of Life* E.O. Wilson gives the proper response to such misplaced optimism.

> Such is the extrapolation endpoint of techno-mania applied to the natural world.
>
> The compelling response, in my opinion, is that to travel even partway there would be a dangerous gamble, a single throw of the dice with the future of life on the table. To revive or synthesize the thousands of species needed—probably millions when the still largely unknown microorganisms have been cataloged—and put them together in functioning ecosystems is beyond even the theoretical imagination of existing science. Each species is adapted to particular physical and chemical environments within the habitat. Each species has evolved to fit together with certain other species in ways biologists are only beginning to understand.[12]

Wilson, of course, is speaking here of whole ecosystems, not farms. But, again, farms are simply biotic communities that are integral to the ecosystem in which they exist. Consequently, anywhere agriculture is practiced, it must become part and parcel of the task of restoring the species diversity that is as essential to a healthy farm as it is to a healthy ecosystem.[13]

Industrial agriculture with its specialization, centralization, and uniformity is simply another example of what Wilson calls "mistaken capital investment." We must now redesign agriculture so that it becomes an integral part of restoring the landscape's biodiversity.

And the reason the human resource factor on farms is important to that task is that such restoration is not likely to be accomplished without caring people on the land. As Berry has reminded us, "there is a limit beyond which machines and chemicals cannot replace people; there is a limit beyond which mechanical or economic efficiency cannot replace care."[14]

Technological Transformations

A third effect likely to result from this new level of industrialization is a further reinforcement of authoritarian technologies. Lewis Mumford, arguably one of America's most important social critics, reminds us that from Neolithic times to the present two technologies have "recurrently existed side by side"—one authoritarian, the other democratic. The former, while powerful, is "inherently unstable"; the latter, while relatively weak, is "resourceful and durable."

Mumford reminds us that democratic technologies usually consist of "the small scale method of production, resting mainly on human skill . . . remaining under the active direction of the craftsman or farmer, each group developing its own gifts, through appropriate arts and social ceremonies, as well as making discreet use of wide diffusion and its modest demands . . . [and has] great powers of adaptation and recuperation." Authoritarian technologies, on the other hand, tend to be large in scale and concentrate power in the hands of the few. They rest mainly on technocratic inventions and scientific discoveries. They are generally under the direction of centralized management, usually exploiting the gifts of nature to suit the purposes of management. Because of its centralization and insatiable demands, authoritarian technology has little power of adaptation or recuperation.[15]

The Next Twenty-Five Years

The overall effect of forcing farms to become industrial complexes that utilize authoritarian technologies will be to change fundamentally the social and biophysical character of both rural and urban communities.

Given these factors, it is not hard to imagine what the landscape of the future may look like in another twenty-five years.

The most extreme outcome might be the one proposed by Steven Blank, a University of California agriculture economist. In his thought-provoking book, *The End of Agriculture in the American Portfolio*, Blank argues that since U.S. farmers cannot compete in the production of bulk commodities with farmers in other parts of the world having significantly lower land and labor costs, we should just get out of the farming business altogether and concentrate our labor on "higher value-producing activities."[16] While others have pointed out the many flaws in Blank's thesis,[17] the market logic of his argument remains persuasive, given the consolidation of the industrial food system and our naive contention that all we need from farmers is the production of cheap raw materials. Fortunately, every urban consumer with whom I have discussed Blank's vision has been horrified, and clearly expressed the view that his vision is simply not an acceptable future. But even if we stop short of Blank's bleak vision, current trends can take us in just ten or fifteen years to a future that may be equally unacceptable to most citizens.

Here is the future I see emerging if present trends continue. A few large industrial farm complexes, managed in accordance with the business interests of five or six consolidated retail firms—perhaps only one of them an American firm (Wal-Mart)[18]—will mass produce a few crop and livestock commodities. These commodities will be produced in intense concentrations, using powerful technologies to achieve specific biological traits, introduced to accomplish the goal of providing the firm with an ample supply of cheap, uniform, raw materials from which to manufacture a variety of food products. These firms will do whatever it takes—to the food and to the body politic[19]—to gain the largest possible share of the consumer's food dollar. Green and purple catsup, chocolate-covered french fries, and inane nutrition advice filtered through the powerful lobbies of the food industry might be only the beginning.

Whatever environmental standards citizens may impose will be adhered to only by means of expensive, draconian regulations. Rural communities will essentially disappear. Trailer parks to house cheap labor will dot the countryside. Riparian buffer strips, grassed waterways, shelterbelts, and terraces will be demolished to make way for equipment that is even larger than the new eighty-foot-wide planters introduced last year. Recreational areas, habitat for wildlife, and wildlife corridors will give way to miles and miles of the continuous expanse of a single crop.

Urban residents will likely have to drive long distances to find suitable recreational or sporting areas. The environmental and human health consequences of imposing high concentrations of single species of crops and animals on the landscape will require an increasing array of powerful technologies to maintain the health of humans and other species. Unpleasant odors—always nature's way of issuing a warning of possible danger—will make it increasingly unpleasant for humans to coexist with agriculture.

And a few parts of rural America, especially places that abound in remote scenic beauty, will become playgrounds for the idle rich.

Meanwhile, global absentee "farmers" will continue to acquire land in developing countries where land and labor costs are cheap in order to produce cheap commodities for export to global consolidated food firms, depriving local populations of the opportunity to access land to feed themselves. This will add another dimension of vulnerability to the industrialized global food system. The social unrest that always accompanies such disparities will continue to destabilize human communities.

The lessons we have learned from evolutionary biology, ecological economics, and the history of social movements suggest that the new industrialized food system described here must, of necessity, be very short-lived on any evolutionary timescale. Evolutionary biology reminds us that population explosions of any species inevitably transform it into a "plague species" that crashes in order to bring it back into ecological balance with the community of species on which it ultimately depends. Very large industrial farms, devoted to producing large quantities of a single species, fit the definition of a plague species exactly.

Ecological economics teaches us that human economies are ultimately subsystems of the natural economy. Natural ecosystems provide the natural resources that fuel human economies, and natural sinks recycle the waste generated by human economies. As these natural systems become degraded or overloaded, human economies begin to fail.

And the history of social movements teaches us that without a modicum of egalitarian rule, societies become destabilized. Such unreliable social environments are almost always costly to the entire biotic community. In today's political and social climate this could be an urgent matter. David Orr reminds us, for example, of some of the vulnerabilities and costs our industrial agriculture system may incur given our current social unrest. Orr suggests that, among other things, a centralized agriculture is not easily protected from terrorist attack.

> A society fed by a few megafarms is far more vulnerable to many kinds of disruption than one with many smaller and widely dispersed farms. One that relies on long-distance transport of essential materials must guard every supply line, but the military capability to do so becomes yet another source of vulnerability and ecological cost. In short, no society that relies on distant sources of food, energy, and materials or heroic feats of technology can be secured indefinitely. . . . An ecological view of security would lead us to rebuild family farms, local enterprises, community prosperity, and regional economies, and to invest in regeneration of natural capital.[20]

Finally, another discomforting analysis of our industrialized systems has been proposed by Charles Perrow in his enlightening book *Normal Accidents*. Perrow argues that as any system becomes increasingly complex and more tightly coupled, normal accidents, which inevitably take place in *any* system, become catastrophes.[21] If Perrow is correct then we should expect an increase in food-related catastrophes in our highly complex, tightly coupled, industrial food system.

Another Vision

But the future I have just described is not inevitable. There are, in fact, encouraging transformations already taking place that suggest that an alternative future may not only be feasible, but highly likely. Aside from the vulnerabilities of the industrial model just described, there are ecological and economic opportunities on the horizon that will clearly give a comparative advantage to a food and farming system that will look very different from the bifurcated system currently emerging. What are some of those opportunities and how might they shape the food and farming system over the next twenty-five years?

New Market Climate

First, a new market climate is emerging that may change how we produce our food. The new market climate, especially where food is concerned, consists of three distinct elements: (1) a "conversational" marketplace; (2) trustworthy and authentic producers and products; (3) a "close to home" connection.

The concept of markets as conversations is imaginatively described in a recent book written by a team of four authors, including two who are on the management team of Sun Microsystems. The book is titled *The Cluetrain Manifesto*.[22] Markets, these authors contend, are undergoing a major change as we enter the twenty-first century. During most of the twentieth century markets consisted of "broadcast" information. To market a product, one published a Sears Roebuck catalog, bought advertising in newspapers or magazines, purchased a spot on the radio, or bought airtime for an advertisement on prime-time television. Marketing meant one-way communication.

The authors of *Cluetrain* argue that the broadcast era is over, that twenty-first-century customers grew up on the Internet and consequently are no longer receptive to having information broadcast to them. They are used to having a conversation about everything—including the products they buy and the food they eat. Therefore, anyone who does not provide an opportunity for customers to have a conversation about what they are selling will simply not be in the market for very long. As *Cluetrain* reminds us, customers are not "seats or eyeballs or end users or consumers," they are human beings whose reach exceeds *your* grasp.

What this analysis of the market of the twenty-first century tells us is that people increasingly will want to have relationships as part of their purchasing experience. Consequently, food marketers of the future who do not provide an opportunity for food customers to experience the story behind the food they buy are not likely to be in that market for very long. When food customers go to the farmers' market or buy from their local CSA, they are buying a relationship as much as a food product—this is the special magic behind today's direct market success. The late Ken Taylor, my good friend and the founder of the Minnesota Food Association, used to express relationship marketing in graphic terms. "People who live in urban communities for the most part don't like to get their hands dirty," he said, "but they surely want to shake the hand of someone that does."

What are the implications of this transformation in the marketplace for the future of agrarian agriculture? In the first place, this new development clearly gives the comparative advantage to precisely those farmers who are most threatened in the emerging bifurcated food system. Imagine with me a large number of small and midsize family farmers producing food products using sound conservation practices, providing their animals with the opportunity to live as nature intended, preserving the identity of such food products by processing them in locally owned processing facilities, and making them available in the marketplace with opportunities for consumers to access the entire story of the product's life cycle.

There are numerous successful examples of such marketing relationships already in place. Let me cite just one such example. A company in Great Britain (home of mad cow disease and the recent hoof and mouth disease outbreak that caused beef sales to plummet) created a new line of products called "Greenstuff." The Greenstuff label carries an identification number that allows food customers to use the Internet to access the story behind each product. According to a story in a British newspaper, this marketing system allows eaters to "see their weekend joint when it still had horns and hooves, contentedly grazing on the farm where it was bred. Shoppers who buy Greenstuff products—a branded range of organic Irish beef, lamb and sausages—can log onto an Internet site for a full CV on the animal they are about to eat."[23]

Here is how it works. Farmers who raise their beef in accordance with the Greenstuff standards take video pictures of their operations—farmers standing in the midst of their herds, cows grazing contentedly in their pastures. Each animal is identified; its medical history, the age at which it was butchered, and other vital statistics are published on the website. In the comfort of their homes, food customers can place their order on the Internet for a five-pound beefsteak from a specific animal, raised by a specific farmer. If they wish, they can contact the farmer by phone. Once the order is placed, the beefsteak from the specific animal with its own unique story is delivered to the door by courier.

How does Wal-Mart compete with such marketing relationships? It becomes extremely difficult for large consolidated firms, whose marketing advantage is based on being the lowest cost supplier of an undifferentiated, mass-produced commodity to provide such marketing relationships.

As Michael Porter reminds us, there are two ways to be competitive in a global economy. One is to supply an undifferentiated product at the lowest cost, and the other is "to provide unique and superior value to the buyer in terms of product quality, special features, or after-sales service."[24] And Porter goes on to say that while it is not impossible for the same firm to do both, it proves to be extremely difficult. The comparative advantage consequently goes to the agrarian farmers of the future.

The second change in today's emerging market climate—the desire for trust and authenticity—also offers a comparative advantage to independent, small and midsize producers. Marketing surveys conducted by the Hartmann Group consistently show that one of the attributes that today's discriminating food customers look for in the marketplace is a trusting relationship and the assurance that claims made for a product are authentic. Smaller supply chains that have food provided for them by identifiable farmers whose products can be easily traced to specific farms—and, as we have seen, even to specific animals—can more readily establish trusting relationships and provide information that can be authenticated than can the consolidated supply chains with mass-produced products. Add to that the opportunity to engage in conversations with the farmers concerning the product and the comparative advantage becomes a no-brainer.

The third change in today's emerging market climate—the increasing desire to purchase "close to home"—also gives a comparative advantage to independent family farms. In a March 4, 2002, interview on the

National Public Radio program *Morning Edition,* Fred Crawford, an internationally recognized retail food consultant, remarked that supermarkets, part of a global economy, won't be in business very long if they fail to realize that their markets are local. Once again, it is agrarian farmers in local communities whose products can be traced directly to the farm—even when those products are sold at some distance—and can be more readily identified as "produced close to home" than mass-produced commodities whose traceability is near impossible.

All of these changes in the market climate suggest new opportunities for relationship marketing that can help convert the most vulnerable farmers in the industrial food system into thriving entrepreneurs in the evolving markets of the next several decades. This bodes well for a new kind of agrarianism.

New Production Paradigm

In addition to the changing market climate, new production strategies that fundamentally redesign farming systems also may present a comparative advantage to the agrarian farmers of the future. These new production strategies evolve out of an ecological—in contrast with a technological—paradigm. Instead of using one-dimensional, single-tactic approaches to solving production problems—an approach that always calls for the invention of yet another technology that farmers must buy—this approach seeks to employ natural systems that are more self-regulatory and synergistic. Using nature as the model, mentor, and measure,[25] these new systems seek to achieve production goals by making use of nature's own free ecosystem services. This natural systems approach to agriculture is, of course, the new design being developed at the Land Institute in Salina, Kansas.

One example of how such systems may change the fundamental parameters of agriculture in the future is the integrated duck/rice system, developed by Takao Furuno, a farmer in southern Japan.[26] Instead of producing rice in a monoculture, dependent on fertilizers and pesticides to achieve acceptable yields, Mr. Furuno developed an elegant, complex, species-interdependent system that has increased his rice yields while producing a full range of other food products, without relying on any outside crop inputs.

Here is how it works. Right after Mr. Furuno sets his rice seedlings out into his flooded rice paddies, he puts a gaggle of young ducklings into the paddies. The ducklings immediately start to feed on insects that normally attack young rice plants. Mr. Furuno then introduces loaches, a variety of fish that is easily cultivated and produces a delicious meat product. He also introduces azolla, normally considered a "paddy weed." The azolla fixes nitrogen but also serves as food for the fish and the ducks. In this way Mr. Furuno has developed a highly synergistic farming system. The ducks feed on the insects and, later, the golden snails that also attack rice plants. Since the ducks and fish feed on the azolla, its growth is kept sufficiently under control so it does not compete with the growing rice, but serves as a source of nitrogen. The nitrogen from the azolla, plus the droppings from the ducks and fish, provides all of the nutrients needed for the rice.

Mr. Furuno also grows figs on the periphery of his rice paddies, supplying him with fruit. He then rotates his integrated rice/duck crop with a crop of vegetables and wheat. He also harvests duck eggs that he markets along with the rice, fish and duck meat, vegetables, wheat, and figs. Mr. Furuno's productivity is also enhanced by the fact that his rice yields in this system exceed the rice yields of industrial rice systems by 20 to 50 percent. This natural systems design makes Mr. Furuno's six-acre farm in Japan one of the most productive in the world. According to conversations he has had with modern monocrop rice growers in Texas, the gross income from Mr. Furuno's six-acre farm in Japan slightly exceeds the gross income of a typical six-hundred-acre rice farm in Texas.[27]

The concept behind Mr. Furuno's design is simple, yet profound. The concept, he writes, "is to produce a variety of products within a limited space to achieve maximum overall productivity. But this does not consist of merely assembling all of the components; it consists of allowing all components to influence each other positively in a relationship of symbiotic production." Farmers, largely without the help of experts, have been developing such elegant synergistic systems throughout the world in recent decades.[28]

These highly productive, redesigned systems, are, once again, most compatible with smaller-scale, independently owned farms. Such complex systems do not lend themselves well to large-scale, highly centralized operations where farmers are seldom intimately involved with the ecology of their farms. It is therefore difficult to imagine them managing such interdependent systems.

New Public Policies

In addition to new market relationships and redesigned production systems, the new agrarianism of the twenty-first century could be further advanced with new public policies. Current agricultural policies are geared to propping up a failed agricultural system. One can reasonably argue that the only reason the consolidated industrial food system has survived as long as it has is because of the immense political power the industry enlists to protect its interests,[29] which include enormous public subsidies that distort free market signals that might otherwise give the competitive advantage to more agrarian production systems. One simple shift in public policy could make an enormous difference. Directing public support *away* from subsidizing a few bulk commodities—a policy that benefits no one except the consolidated firms that acquire those commodities well below the cost of production—and *toward* the support of public goods that benefit communities could help transform much of agriculture.

We all recognize that farmers produce more than food and fiber. They also provide a wide range of public services. For example, properly managed soils help filter water to improve water quality. Properly managed landscapes provide habitat for wildlife, helping to restore biodiversity and provide recreational space for hunting and fishing. One can reasonably argue that the public ought to pay for such services, provided the farmer conserves and nurtures these resources. Given adequate information and incentives, farmers could provide additional public services that could benefit the entire citizenry—economically, socially, and ecologically. Provided with adequate support, farmers could be in the forefront of providing all of society with clean water, clean air, quality soil, vibrant communities, and a host of other services.

So we do have a choice. We can continue on our current path, insisting that we have too many farmers and that economic efficiency is the sole driver of our future food and farming system, ultimately leaving us with industrial complexes on the American landscape whose sole objective is to produce raw materials for food and fiber as cheaply as possible regardless of the cost to the environment or human communities. Or we can invest in new marketing systems that enable farmers to produce more value and retain more of that value on the farm; empower the research community to develop more productive, diverse, biologically rich production systems that mimic nature and are more profitable for farmers and less harmful to the environment; and develop new public policies that begin the transition to a new food and farming future.

But, once again, this will require a particular kind of farmer with a particular kind of farm. Of all the millions of words that have been written about agriculture since the publication of *Unsettling*, none have described what we need more eloquently than the particulars outlined by Wendell Berry in another of his poignant works written over a decade after *Unsettling*. I find myself, again and again, going back to this simple description of what we need: "if agriculture is to remain productive, it must preserve the land, and the fertility and ecological health of the land; the land, that is, must be used well. A further requirement, therefore, is that if the land is to be used well, the people who use it must know it well, must have time to use it well, and must be able to afford to use it well. Nothing that has happened in the agricultural revolution of the last fifty years has disproved or invalidated these requirements, though everything that has happened has ignored or defied them."[30]

NOTES

1. Calvin Beale, "Salient Features of the Demography of American Agriculture," in *The Demography of Rural Life* (publication 64), ed. David Brown et al. (University Park, Pa.: Northeast Regional Center for Rural Development, 1993).

2. Willard W. Cochrane, "A Food and Agriculture Policy for the Twenty-First Century," June 21, 1999 (unpublished paper available from the author).

3. Frederick H. Buttel, "Ideology and Agricultural Technology in the Late

Twentieth Century: Biotechnology as Symbol and Substance," *Agriculture and Human Values* 10, no. 2 (spring 1993).

4. Cochrane, "A Food and Agriculture Policy."

5. Ibid. These are the farms that are the most vulnerable since they are too small to access markets with consolidated firms and have little opportunity to do direct marketing.

6. See the work of William Heffernan, who has tracked the consolidation in the agriculture and food industry for the past twenty-five years. His most recent work (coauthored with Mary Hendrickson and others), "Consolidation in Food Retailing and Dairy: Implications for Farmers and Consumers in a Global Food System," January 8, 2001, is available from the National Farmers Union.

7. W.C. Lowdermilk, "Conquest of the Land through Seven Thousand Years," USDA, Soil Conservation Service, Agriculture Information Bulletin no. 99, 1953.

8. Michael Boehlje, "Structural Changes in the Agricultural Industries: How Do We Measure, Analyze and Understand Them?" *American Journal of Agricultural Economics* 81, no. 5 (December 1999).

9. *Time,* "Arkansas Pecking Order," October 26, 1992.

10. See, for example, David Tilman's work: "The Greening of the Green Revolution," *Nature* 396 (November 19, 1998); "Biodiversity and Ecosystem Functioning: Current Knowledge and Future Challenges," *Science* 294 (October 26, 2001).

11. Quoted in William Greider, "The Last Farm Crisis," *The Nation*, November 20, 2000.

12. Edward O. Wilson, *The Future of Life* (New York: Knopf, 2002).

13. For an excellent analysis of how and why agriculture must transform itself from a fossil fuel-based system to a species interaction system, see Masae Shiyomi and Hiroshi Koizumi, *Structure and Function in Agroecosystem Design and Management* (New York: CRC Press, 2001).

14. Wendell Berry, *Another Turn of the Crank* (Washington, D.C.:Counterpoint Press, 1995).

15. Lewis Mumford, "Authoritarian and Democratic Technics," in *Questioning Technology*, ed. John Zerzan and Alice Carnes (Philadelphia, Pa.: New Society Publications, 1991).

16. Steven C. Blank, *The End of Agriculture in the American Portfolio* (Westport, Conn.: Greenwood Publishing, 1998). Other industrial countries are also beginning to presage the end of agriculture in their lands. See, for example, Andrew O'Hagan, *The End of British Farming* (London: Profile Books, 2001).

17. See, for example, Neil Harl, "Review of *The End of Agriculture in the American Portfolio,*" http://www.econ.iastate.edu/faculty/harl/Book_Review.html.

18. Heffernan et al., "Consolidation in Food Retailing."

19. Marion Nestle, *Food Politics* (Berkeley: University of California Press, 2002).

20. David Orr, "The Events of 9-11: A View from the Margin," *Conservation Biology* 16, no. 2 (April 2002).

21. Charles Perrow, *Normal Accidents: Living with High-Risk Technologies* (Princeton, N.J.: Princeton University Press, 1999).

22. Christopher Lacke et al., *The Cluetrain Manifesto* (Boulder, Colo.: Perseus Books Group, 2000).

23. Jo Knowsley, "You Saw the Cow, Now Eat Its Meat," *The Mail on Sunday,* September 2, 2001.

24. Michael E. Porter, *The Competitive Advantage of Nations* (New York: The Free Press, 1990).

25. See Janine M. Benyus, *Biomimicry* (New York: William Morrow, 1997).

26. For a detailed description of Mr. Furuno's system see Takao Furuno, *The Power of Duck* (Tasmania, Australia: Tagaari Publications, 2001).

27. Mr. Furuno, conversation with author, October 21, 2001.

28. See, for example, the surveys done by Jules Pretty and Rachel Hine. http://www2.essex.ac.uk/ces/ResearchProgrammes/SAFEW47casessusag.htm

29. Nestle, *Food Politics.*

30. Wendell Berry, *What Are People For?* (San Francisco: North Point Press, 1990).

7

Globalization and the War against Farmers and the Land

Vandana Shiva

I had trained as a quantum physicist, expecting to spend a lifetime solving puzzles in quantum theory. Instead, I have spent the past two decades solving puzzles in agriculture. Why did the seeds of the Green Revolution, which brought Norman Borlaug the Nobel Peace Prize, also become seeds of war in the Indian Punjab during the 1980s, within two decades of their introduction? Why were Indian peasants being systematically pushed into debt and penury by industrial agriculture, which was supposed to create prosperity for rural communities? Why did low-productivity monocultures pass as high-productivity systems even though they depended on high inputs and had lower outputs than biodiverse farms?

1984 was a dramatic and tragic year for India. It was the year the violence in Punjab, which claimed thirty thousand lives, reached its climax, with the army entering the Golden Temple in June to wipe out the extremists, and with Prime Minister Indira Gandhi being assassinated in retaliation. It was also the year of the Bhopal tragedy in which three thousand people were killed by a leak from a Union Carbide pesticide plant. Thousands were crippled for life. And in the summer of 1984, Karnataka experienced drought, in spite of normal rainfall. The drought was created by the Green Revolution's "miracle" seeds that reduced biomass, creating a fodder famine and depleting organic matter in the soil. Soils, starved of organic matter that the tall indigenous sorghum varieties provided, could no longer store the rainfall as soil moisture, and drought and desertification resulted from a short-term, narrow-minded search for high yields of grain. It was during the drought that an old farmer made the links between native seeds and ecological security. His wisdom sowed in me the passion to save seeds.

During these turbulent times I tried to understand how we had landed in such a mess with the dominant paradigm of agriculture. The official literature failed to address the puzzles—it was more like propaganda. My search led me to Wendell Berry's writing. I could see that what was happening to India had also happened to agriculture in the United States. The unsettling of India was repeating the pattern of the unsettling of America.

AGRICULTURE AS WAR

Agriculture based on diversity, decentralization, and improved small farm productivity through ecological methods is a women-centered, nature-friendly agriculture. Here knowledge is shared, other species and plants are viewed as kin rather than "property," and sustainability is based on the renewal and regeneration of biodiversity and species richness on farms that provide internal inputs as alternatives to external inputs of synthetic fertilizers and chemicals. In this paradigm, there is no place for monocultures of genetically engineered crops and corporate monopolies on seeds.

Industrial agriculture, on the other hand, has become a war against ecosystems. It is based on the instruments of war and the logic of war, and it has warlike consequences. The chemicals on which industrial agriculture is based were originally designed for chemical warfare. That is why industrialists turned Bhopal into a war zone. That is why corporate agriculture converts our farms into war zones.

Monocultures and monopolies symbolize a masculinization of agriculture. The aggressive, competitive mentality underlying military-industrial agriculture is evident in the names given to herbicides that destroy the economic basis of survival for the poorest women in the rural areas of the

Third World. Monsanto's herbicides are called "Roundup," "Machete," and "Lasso." American Home Products, which has merged with Monsanto, calls its herbicides "Pentagon," "Prowl," "Lightning," "Assert," and "Avenge." This is the language of war, not sustainability. Sustainability is based on peace with the earth.[1]

What we may fail to see is that the violence in agriculture is linked to terrorism as we have seen it in the Indian Punjab and the plains of the United States. In the 1980s, the farm crisis fueled violent Sikh nationalism, as unemployed and angry youth took up arms exported by the same global powers that had destroyed Indian agriculture and who looked on India as a market for their overpriced, non-essential, often hazardous products and technologies. Likewise, the Oklahoma City bombing was linked to a national farm crisis, as evident in the growing dispossession and frustration of American family farmers who increasingly turned to a gospel of violence and hatred being promoted by Christian militias. As Joel Dyer says in *Harvest of Rage: Why Oklahoma City Is Only the Beginning*,

> America's innocence lay in the rubble of the Murrah building as surely as the crumpled bodies of the victims. The deadly Oklahoma City bomb was just the first shot in the collective suicide of the nation. Some Americans—some of them our neighbors—have declared war on the powers that be, and those of us who stand unknowingly in between these warring factions are paying the price. And we will continue to pay the price—one building, one pipe bomb, one burned-down church at a time—until we come to understand, first, that the nation is holding a loaded gun to its head and, second, why so many among us are struggling to pull the trigger.[2]

As we destroy the ecological and social community in our countryside—all in the name of corporate profit—we increase the threshold of violence and decrease our capacity for compassion. It is in the daily, responsible interaction between species that we learn our best lessons in diversity and democracy.

The Green Revolution package was built on the displacement of genetic diversity at two levels. First, mixtures and rotation of diverse crops like wheat, maize, millets, legumes, and oilseeds were replaced by large acreage monocultures of wheat and rice. Second, the introduced wheat and rice varieties came from a very narrow genetic base, compared to the high genetic variability in the populations of traditional wheat and rice plants. When these high-yield seeds replace native seed systems, species diversity and resilience are lost irreversibly. The destruction of diversity and the creation of uniformity simultaneously entail ecological instability and vulnerability.

As in the rest of India, indigenous agriculture in Punjab was based on diversity. Nonfood crops included indigo, sugarcane, cotton, hemp, asafetida, and oilseeds. The horticultural crops included guavas, dates, mangoes, limes, lemons, peaches, apricots, figs, pomegranates, plums, oranges, mulberries, grapes, almonds, melons, apples, beans, cucumbers, carrots, and turnips. The uncultivated areas were covered by date palm, wild palm, willow, acacia, sissoo (rosewood), and other trees and shrubs. The millets, called "minor cereals" (because they are so diverse, not because they are insignificant crops), occupied the largest area under cultivation in Punjab. "Kutki," the little millet (*Panicum miliare*), "jowar" (*Sorghum vulgare*), "ragi" (*Eleusine coracana*), and "bajra" or bulrush millet (*Pennisetum typhoides*) were the main millets cultivated in Punjab, covering 43 percent of the area. Besides these, there were uncultivated or wild varieties of millet like "shama" (*Panicum hydaspicum*), *Cenchrus echinatus*, and *Pennisetum cenchroides*. In addition to these, were the more well-known cereals "makki" (maize) and wheat. Closely related to the millets were lesser-known crops like amaranth, of which Punjab had a rich diversity. "Sil" or "mawal" (*Celosia cristata*) grew both cultivated or wild. "Gauhar," "sawal" sil, "bhabri," "savalana," "batu," and "chaulai" were the different names given to the common amaranth (*Amaranthus paniculatus*). An important crop used for green leafy vegetable was "bathua" (*Chenopodium album*). The legumes of Punjab included "moth-safaid" (*Cyamopsis psoralioides*), "channa" or chickpea (*Cicer arietinum*), "bhat" (*Glycine soya*), "urd" and "mash" (*Phaseolus mungo*), "lobiya" (*Vigna catiang*), "kulat" (*Dolichos biflorus*), "kharnab-mibti" (*Ceratonia siliqua*). The oilseeds included "til" or sesame (*Sesamum indicum*), groundnuts (*Arachis hypogaea*), "alsi" or linseed (*Linum usitatissimum*), and "sarson" or mustard (*Brassica nigra*). The cereals, legumes, and oilseeds were grown in various mixtures and rotations.[3] This diversity was crucial not only for ecological reasons, but also because it assured a food crop for farmers unable to afford expensive industrial technologies. But as diversity gave way to monocultures, cultures of peace and sharing gave way to cultures of violence. It was not peace but war that was the legacy of the Green Revolution.

The War against Farmers

Industrial agriculture is also translating into economic warfare against the poor. Hunger has grown in the Third World in direct proportion to the spread of industrial agriculture and the globalization of trade in staple foods. This is no accident. Industrial agriculture is an efficient system for robbing farmers of wealth and pushing them into debt and dispossession. The costly seeds, chemicals, and machinery that replaced the farm's internal resources were originally supported through subsidies. Today, they are obtained by borrowing from the same agents who sell the pesticides and seeds. A new phenomenon of corporate feudalism is emerging, as global seed and agrichemical corporations combine with the local feudal power of landlords and moneylenders to trap innocent peasants into unpayable debt. More than twenty thousand Indian peasants have committed suicide since the seed and agriculture sector was opened up to global corporations. The worst suicides have occurred in Warangal in Andhra Pradesh.

Of late some farmers in Rentichintala and surrounding areas like Gurazala have sold their kidneys in order to clear their outstanding debts with the *pawn*brokers. The farmers who sold their kidneys from Rentichintala Mandal are:

1. Durgyampudi Chinna Venkat Reddy
2. Dirsinals Narsi Reddy
3. Bobba Venkat Reddy
4. Siddhavarpu Poli Reddy
5. Peram Lacchi Reddy
6. Kancharla Krishna

7. Narmala Krishna
8. Golle Ramaswami
9. Thai Narasaiah[4]

Bhatinda in Punjab has the second highest rate of farm suicides. Over the past few decades, costly pesticide use has increased there by 6000 percent. As we can see, the growth in seed and chemical markets for global corporations is based on taking the lives and livelihoods of peasants of the Third World.

All manner of false claims are made by the seed corporations to entice and entrap farmers. While the rhetoric of the "Green Revolution" and genetic engineering is the removal of hunger, the reality is that high-cost, high external input agriculture creates hunger by leaving nothing in rural households. Peasants must sell all they produce in order to pay back debts. That is why the producers of food are going hungry themselves. And 65 million tonnes of grain rot in storage because people have been robbed of their entitlements and purchasing power to access the food they have produced.

THE SYSTEMIC ROOTS OF HUNGER IN FREE TRADE AND INDUSTRIAL FARMING

In 1942 more than 3 million people died in Bengal and Orissa due to starvation. Nobel Prize–winning economist Amartya Sen showed that it was not lack of food but lack of food entitlements and food rights that caused starvation deaths. The Indian Constitution guarantees the right to life, and the right to food is at the heart of the right to life. The state has a related duty to ensure that no one goes hungry. Food-for-work programs, the public distribution system, price regulation, and anti-hoarding measures have been diverse policy components ensuring that people's food entitlements are protected. Ensuring food rights in a poor country like India requires the protection of livelihoods, the promotion of low-cost sustainable agriculture, and localization and decentralization of food distribution networks to reduce costs and waste.

All elements of India's food security policy are today being dismantled under pressure from the World Bank and the World Trade Organization (WTO). Starvation is the inevitable result of policies promoting sudden withdrawal of the protective role of the state and reckless dependence on markets to bring food to the poor who have no purchasing power. For the first time since the Great Bengal Famine of 1942, created by British imperial free trade policies, famine and starvation have returned to India. During the monsoon season of 2001, both the Indian Parliament and the Indian Supreme Court had to intervene in the government's trade liberalization policies, which were causing starvation deaths in India.

Global trade agreements and international financial institutions are preventing the government from supporting the poor in their access to adequate and nutritious food. They are promoting the diversion of subsidies from people to corporations. Exports increase while people starve. Corporations are subsidized while people's food subsidies are withdrawn. This is how globalization is causing hunger and starvation in the Third World.

It is the trading giants like Pepsi and Cargill who have benefited from the redirection of food subsidies benefiting the poor. Pepsi Foods, a Pepsi subsidiary involved in contract farming, exported one hundred thousand tonnes of rice from India during 2002 for a profit of 12.2 million rupees while people in India faced starvation. Meanwhile Cargill exported tonnes of wheat in 2001 and planned to procure much more during the 2002 harvest. Trade liberalization is a recipe for starving the poor to feed the corporations. While the World Bank and International Monetary Fund remove subsidies from food for the poor, they encourage subsidies to grain giants like Cargill and Pepsi for exporting food grain.

Hunger is also being caused by globalization, which is removing the floor from prices of agricultural commodities under the dual pressure of export subsidies and competition between producing countries. Thus the coffee trade has increased from $40 billion to $70 billion over the past few years, even as the price coffee growers receive has dropped from $9 billion to $5.5 billion. Coffee growers in Latin America and Africa are starving. The case of coffee shows that growth in global trade and commerce does not translate into growth for poor producers. It can, in fact, worsen their poverty.

In December 1999 the United States forced India to remove import restrictions (quantitative restrictions or QRs) on imports. QRs are legitimate controls that a country can choose to impose by restricting the quantity of imports and exports to protect its domestic producers. Imports have been restricted by countries on various grounds for environmental and ethical reasons or reasons of public order. India has been maintaining QRs on several items, as when the government places items on the restricted list of imports—the canalized list, which channels imported commodities through public sector companies—and lists items for which a special import license (SIL) is required.

The control of imports/exports through QRs helps to protect livelihoods, especially those in informal economic sectors where a large majority of our population and those who are largely self-employed work. In the light of huge inequalities that prevail between and within countries, tariffs alone cannot provide adequate safeguards against the uncontrolled entry of foreign goods. Even so, the entire economy of India was handed over to the United States in a secret deal to remove import restrictions on 714 items by 1 April 2000 and 715 items by 1 April 2001.

Artificially cheap subsidized products like soy oil soon started to flood the market. Imports of soybean oil have increased from 236,000 tonnes in 1997–98 to 800,000 in 1998–99. In 2000 it was estimated that 5.5 million tonnes of foreign palm oil, soyabean oil, and animal fat would be imported. The mustard produced by our farmers, which was selling at 2,000 rupees per quintal in 1999, is today not even selling for 900 rupees per quintal. The production of mustard seeds has fallen by 65 percent, and over 60 percent of small oil mills and ghanis (very small, cold-pressed mills) have been closed down, leaving many people unemployed.

As a result of unfair trading practices legalized by the WTO, India's agricultural imports have gone up from 50 billion rupees in 1995 to over 200 billion rupees in 1999–2000, a 400 percent increase in imports. Similarly, soybean and palm oil have flooded the Indian market, destroying the domestic edible oil economy based on coconut, mustard, groundnuts, and sesame. In response, coconut farmers in Kerala blockaded the Cochin harbor, and groundnut farmers in Sirsi, Karnataka, and soybean farmers in Multai (a district of Madhya Pradesh) protested, drawing gunfire.

What we can now see is that price collapse is not a linear mechanical phenomenon dependent on percentage of imports. It is more appropriately described in terms of a nonlinear perturbation in a complex system that can slide the system into chaos and disintegration. In the case of products with low import levels, removal of import restrictions has sent the domestic prices into a downward spin, leaving the producers in crisis and the agricultural economy in shambles. Meanwhile a new farm bill in the United States raises agricultural subsidies over the next six years to $180 billion, an increase of $72.5 billion over current subsidies. This will artificially bring down prices worldwide, destroying millions of peasant livelihoods. It is not efficiency but subsidies that are lowering prices of farm commodities.

This is why the global movement for food rights and farmer's rights demands that food and agriculture not be governed by rules of free trade. Countries have a right to restrict imports. Food and livelihood security requires that purchase prices reflect the full cost of production, including the sustenance and fair returns provided to farmers for their labor.

Using Hunger to Market Biotech

The crisis of farming created by globalization is now being used by the biotechnology industry to market genetically modified (GM) seeds in the Third World. The World Food Summit meeting in Rome was supposed to address the worst human rights violation of our time—the denial of the right to food to millions. But delegates found soccer more important than hunger. Silvio Berlusconi, the Italian prime minister and chair of the meeting, wrapped up the so-called summit a full two hours before schedule so that delegates could watch the World Cup. No serious commitments were made, nor was any serious analysis devoted to addressing the mounting international crisis of hunger and malnutrition.

While the summit was a total failure in addressing the hunger issue, it did become a launching pad for the biotechnology industry. The hunger for food was neglected. The hunger for profit was fully attended to. It was used to put the stamp of approval on genetically engineered seeds and crops that have been at the center of worldwide controversy over the past decade.

As is becoming the trend, the World Food Summit was not negotiated—a text was ready before the leaders arrived. The leaders came only from the southern hemisphere. The rich countries' leaders were conspicuous by their absence, but the United States government was conspicuous by its influence. USDA Secretary Ann Veneman, who used to be with Calgene, now a company under Monsanto, held press conferences to announce how biotechnology would save people and the rainforests. (An American journalist who interviewed me informed me that the current United States government is in fact a "Monsanto administration." Defense Secretary Donald Rumsfeld used to be president of Searle, which merged with Monsanto, and Attorney General John Ashcroft has received campaign funds from Monsanto.)

While no financial commitment was made to alleviate hunger, the head of USAID announced $100 million in biotechnology aid to Third World countries over the next ten years for transfer of biotechnology. The offer comes tied to trade and commercial interests. It would not be a surprise if the poor countries who have been resisting genetic engineering so far now open their doors to it. Indeed the World Food Summit seemed more like a sale show for the biotech industry than a serious gathering of leaders collectively seeking ways and making commitments to address the biggest human rights disaster of our times: more than 1 billion people going hungry in a world abounding in food and wealth.[5]

The Earth Summit in Johannesburg in 2002, organized ten years after the Rio Summit, was also reduced to a marketplace for pushing biotech on Africa. Many southern African countries face drought and famine under the joint impact of climate change and structural adjustment. The Zambian and Zimbabwean governments rejected GM food aid. Hundreds of African farmers and government representatives condemned the U.S. pressure to force GM-contaminated food aid upon them. As a civil society representative stated,

> We, African Civil Society groups, participants to the World Summit on Sustainable Development, composed of more than 45 African countries, join hands with the Zambian and Zimbabwean govern-

ments and their people in rejecting GE contaminated food for our starving brothers and sisters:

1. We refuse to be used as the dumping ground for contaminated food, rejected by the Northern countries; and we are enraged by the emotional blackmail of vulnerable people in need, being used in this way.
2. The starvation period is anticipated to begin early in 2003, so that there is enough time to source uncontaminated food.
3. There is enough food in the rest of Africa (already offered by Tanzania and Uganda) to provide food for the drought areas.
4. Our response is to strengthen solidarity and self-reliance within Africa, in the face of this next wave of colonization, through GE technologies, which aim to control our agricultural systems through the manipulation of seed by corporations.
5. We will stand together in preventing our continent from being contaminated by genetically engineered crops, as a responsibility to our future generations.

When Colin Powell, representing President Bush, kept insisting in the closing plenary session that African countries should import GM food from the United States, he was heckled by both nongovernmental organizations (NGOs) and governments. African farmers had come to Johannesburg with genuine alternatives—small-scale, indigenous-based farming methods predicated on farmers' rights to land, water, and seed. They were not heard. Instead, the globalization and corporatization of agriculture was falsely justified on claims of the higher productivity of large industrial farms.

The Myth of Productivity

A frequent argument used in promoting industrial agriculture, formerly with the Green Revolution and now with genetic engineering, is that only industrial agriculture and industrial breeding can keep up with increased food productivity for feeding a growing population. However, if we are to increase the number of mouths we can feed we must promote more efficient resource use so that the same resources can feed more people. Since resources, not labor, are the limiting factor in food production, it is resource productivity, not labor productivity, that is the relevant measure. A sixty-fold decrease in food producing capacity due to compromised ecological/agricultural habitats and the destruction of biodiversity is not an efficient strategy for feeding a growing world.

Further, since food security is based on food entitlements, and entitlements in peasant societies are based on livelihoods and work, increase in labor productivity based on the displacement of farmers from rural to urban areas cannot reduce hunger. It decreases food entitlements by destroying livelihoods.

Contrary to the myth of linear, industrial progress, comparative studies of twenty-two rice-growing systems have shown that indigenous systems are more efficient in terms of yields, and in terms of labor use and energy use. Both from the point of view of food productivity and of food entitlements, industrial agriculture is deficient as compared to sustainable farming systems based on diversity and internal inputs. While the partial productivity of industrial agriculture based on resource and energy intensification (high amounts of fossil fuel consumption) is non-sustainable both because of environmental externalities and because of livelihood destruction, biodiversity intensification leads to an increase in food production by increasing resource productivity as well as increasing employment, all the while improving the health and stability of food-producing environments.

Not only is the measure of productivity of industrial agriculture partial because all inputs, including resource and energy inputs, are not taken into account, it is also partial because not all outputs are taken into account.

Ecological agriculture is based on mixed and rotational cropping and the production of a diversity of crops. The polycultures of traditional agricultural systems have evolved because more yield can be harvested from a given area planted with diverse crops than from an equivalent area consisting of separate patches of monocultures. For example, in plantings of sorghum and pigeon pea mixtures, one hectare will produce the same yield as 0.94 hectares of sorghum monocultures and 0.68 hectares of pigeon pea monoculture combined. Thus one hectare of polyculture produces what 1.62 hectares of monoculture can produce. This is called the land equivalent ratio (LER).

Increased land-use efficiency and higher LERs have been reported for polycultures of millet/groundnut (1.26); maize/bean (1.38); millet/sorghum (1.53); maize/pigeon pea (1.85); maize/cocoyan/sweet potato (2.08); cassava/maize/groundnut (2.51). The monocultures of the Green Revolution thus actually reduced food yields per acre when compared with mixtures of diverse crops. This falsifies the argument often made that chemically intensive agriculture and genetic engineering will save biodiversity by releasing land from food production. In fact, since monocultures require more land, biodiversity is destroyed twice over—once on the farm, and then on the additional acreage required to produce the outputs a monoculture has displaced. Not only is the productivity measure distorted by ignoring resource inputs (focusing only on labor), it is also distorted by looking at a single and partial output rather than the total food output.

A myth promoted by the one-dimensional monoculture paradigm is that biodiversity reduces yields and productivity while monocultures increase yields and productivity. However, since yields and productivity are theoretically constructed terms, they change according to the context. Yield usually refers to production per unit area of a single crop. Planting only one crop in the entire field as a monoculture will of course increase its yield. Planting multiple crops in a mixture will have low yields of individual crops, but will have high total output of food.

The Mayan peasants in the Mexican state of Chiapas are characterized as unproductive because they produce only two tonnes of corn per acre. However, the overall food output is twenty tonnes per acre. In the terraced fields of the high Himalayas, women peasants grow jhangora (barnyard millet), marsha (amaranth), tur (pigeon pea), urad (black gram), gahat (horse gram), soybean (glysine max), bhat (glysine soya), rayans (rice bean), swanta (cowpea), and kodo (finger millet) in mixtures and rotations. The total output, even in bad years, is six times more than industrially farmed rice monocultures.

The work of the Research Foundation for Science, Technology, and Ecology (based in New Delhi) has shown that farm incomes can increase threefold by giving up chemicals and using internal inputs produced by on-farm biodiversity, including straw, animal manure, and other by-products. Examples of how biodiversity promotes plentiful and stable yields abound.

- Indigenous farmers of the Andes grow more than three thousand varieties of potato.
- In Papua New Guinea, as many as five thousand varieties of sweet potato are under cultivation, with more than twenty varieties grown in a single garden.
- In Java, small farmers cultivate 607 species in their home gardens, with an overall species diversity comparable to a deciduous tropical forest.
- In sub-Saharan Africa, women cultivate as many as 120 different plants in the spaces left alongside the cash crops.
- A single home garden in Thailand has more than 230 species.
- African home gardens have more than 60 species of trees.
- Rural families in the Congo eat leaves from more than 50 different species of trees.
- A study in eastern Nigeria found that home gardens occupying only 2 percent of a household's farmland account for half of the farm's total output.
- Home gardens in Indonesia are estimated to provide more than 20 percent of household income and 40 percent of domestic food supplies.

These food production practices have stood the test of time, whereas the many promises of corporate and biotech companies have hardly been tested, let alone proven, over the long term.

The main argument used for the industrialization of food and corporatization of agriculture is the low productivity of the small farmer. Surely these families on their little plots of land are incapable of meeting the world's need for food! Industrial agriculture claims that it increases yields, hence creating the image that more food is produced per unit acre by industrial means than by the traditional practices of small holders. However, sustainable diversified small-farm systems are actually more productive.

Industrial agriculture productivity is high only in the restricted context of a "part of a part," whether it be the forest or of the farm. For example, "high-yield" plantations pick one tree species among thousands, for yields of one part of the tree (e.g., wood pulp), whereas traditional forestry practices use many parts of many forest species.

"High-yield" Green Revolution cropping patterns select one crop among hundreds, such as wheat, for the use of just one part, the grain, and thus leave out of account the usefulness of the straw for food or compost (and its importance in the maintenance of soil and water stability). These high partial yields do not translate into high total yields, because everything else in the farm system goes to waste. Usually the yield of a single-crop like wheat or maize is singled out and compared to yields of new varieties. This calculation is biased to make the new varieties appear "high yielding" even when, at the systems level, they may not be.

Traditional farming systems are based on mixed and rotational cropping systems of cereals, legumes, and oilseeds with different varieties of each crop, while the Green Revolution package is based on genetically uniform monocultures. No realistic assessments are ever made of the yield of the diverse crop outputs in the mixed and rotational systems.

Productivity is quite different, however, when it is measured in the context of diversity. Biodiversity-based measures of productivity show that small farmers can feed the world. Their multiple yields result in truly high productivity, composed as they are of the multiple yields of diverse species used for diverse purposes. Thus productivity is not lower on smaller units of land: on the contrary, it is higher. In Brazil, the productivity of a farm of up to ten hectares was eighty-five dollars per hectare while the productivity of a five hundred hectare farm was two dollars per hectare. In India, a farm of up to five acres had a productivity of 735 rupees per acre, while a thirty-five-acre farm had a productivity of 346 rupees per acre.

Diversity produces more than monocultures. But monocultures are profitable to industry both for markets and political control. The shift from high-productivity diversity to low-productivity monocultures is possible because the resources destroyed are taken from the poor, while the higher commodity production brings benefits to those with economic power. The polluter does not pay in industrial agriculture. Ironically, and tragically, while the poor go hungry it is the hunger of the poor that is used to justify the agricultural strategies that deepen their hunger.

Diversity has been destroyed in agriculture on the assumption that it is associated with low productivity. This is, however, a false assumption both at the level of individual crops as well as at the level of farming systems. Diverse native varieties are often as high or higher yielding than industrially bred varieties. In addition, diversity in farming systems has a higher output at the total systems level than one-dimensional monocultures.

The measurement of yields and productivity in the Green Revolution as well as in the genetic engineering paradigm is divorced from seeing how the processes of increasing single-species, single-function output affect the processes that sustain the condition for agricultural production, both by reducing species and functional diversity of farming systems as well as by replacing internal inputs provided by biodiversity with hazardous agrichemicals. While these reductionist categories of yield and productivity allow a higher measurement of harvestable yields of single commodities, they measure neither the ecological destruction that affects future yields nor the loss of diverse outputs from biodiversity rich systems.

Productivity in traditional farming practices has always been high if it is remembered that very little external inputs are required. While the Green Revolution has been projected as having increased productivity in the absolute sense, when its expensive resource utilization is taken into account, it has been found to be counterproductive and resource inefficient.

Perhaps one of the most fallacious myths propagated by Green Revolution proponents is the assertion that high-yield varieties have reduced the acreage cultivated, therefore preserving millions of hectares of biodiversity. India's experience tells us that instead of more land being released for conservation, by destroying diversity and multiple uses of land, the industrial breeding system actually increases pressure on the land:

each acre of a monoculture provides a single output, forcing the displaced outputs to be grown on additional acres.

If we focus on land use in the Green Revolution, the industrial breeding strategies increased grain production by 20 percent under good conditions and led to a 100 percent decline of straw. Traditional varieties produce one thousand kilograms/acre of grain and one thousand kilograms/acre of fodder, or two thousand kilograms of grain and two thousand kilograms of fodder on two acres, whereas the industrial strategy only provides twelve hundred kilograms of grain and one thousand kilograms of fodder on two acres, leading to a *decline* of eight hundred kilograms of grain and one thousand kilograms of fodder. Further, the same reductionist logic of industrial breeding also increases the resource use by cattle. Industrial livestock farming consumes three times more biomass than ecological livestock maintenance. Thus industrial livestock breeding would in fact require three times more acres of land for feed. In fact, Europe uses seven times its own area in Third World countries for cattle feed production. For fodder alone (including that used to produce food products for export) the Netherlands appropriates 100,000 to 140,000 square kilometers of arable land, much from the Third World. This is five to seven times the area of agricultural land in the entire country.

The combination of industrial plant breeding and industrial animal breeding therefore *increases the pressure on* land use by a factor of 400 percent while separately increasing output of grain and milk by only a factor of 20 percent. The extra resources used by industrial systems—either by the Green Revolution or the new biotechnologies—could have gone to feed people. Resources wasted amount to the creation of hunger. By being resource wasteful through intensive external inputs, the new biotechnologies create food insecurity and starvation.

What does all this evidence mean in terms of "feeding the world"? It becomes clear that industrial breeding has actually reduced food security by destroying small farms and the small farmers' capacity to produce diverse outputs of nutritious crops. Both from the point of view of food productivity and food entitlements, industrial agriculture is deficient as compared to diversity-based internal input systems. Protecting small farms that conserve biodiversity is thus a food security imperative.[6]

The monoculture of the mind is the disease that blocks the creation of abundance on our small farms.

NAVDANYA: SPREADING NONVIOLENCE IN AGRICULTURE

In 1987 I started Navdanya to save native seeds threatened by erosion of biodiversity and by genetic engineering and patenting. For me the seed became the embodiment and symbol of freedom. As I wrote in 1988 in "The Seed and the Spinning Wheel," "During the first industrial revolution and its associated colonization, Gandhi had transformed the 'primitive' spinning wheel into a living symbol of the struggle for India's freedom and self-determination. The 'primitive' seeds of Third World peasants could well become the symbols of the struggle for freedom and the protection of life in the emerging context of recolonization of the Third World and its living resources."[7]

Over the past fifteen years, more than fifty community seed banks have sprouted for saving and sharing thousands of crop varieties of rice, wheat, oilseeds, millets, legumes, and vegetables across India. Thousands of farmers have given up chemicals and moved to "ahimsic kheti"—nonviolent organic agriculture. Millions have pledged to obey the higher moral law of protecting biodiversity and freely sharing in its bounties and to not cooperate with patent laws that make seed sharing and seed saving an intellectual property crime.

While corporations use falsehoods and coercive power to spread violence in agriculture, we depend on truth and grassroots organization to spread peace and nonviolence. Agriculture has become my experiment with truth and beauty and peace. And for this I draw strength from the boundless energy of the earth and her infinite diversity.

NOTES

1. *Bija—The Seed* (periodical publication of Research Foundation for Science, Technology and Ecology, New Delhi), nos. 21 & 22 (2002).

2. Quoted in Vandana Shiva, "Globalization and Talibanization," in *September 11, 2001: Feminist Perspectives,* ed. Susan Hawthorne and Bronwyn Winter (North Melbourne, Australia: Spinifex Press, 2002).

3. Vandana Shiva, "The Violence of the Green Revolution," in *The Other India* (New Delhi: Research Foundation for Science, Technology and Ecology, 1992), 81–82.

4. Vandana Shiva, et al., *Seeds of Suicide* (New Delhi: Research Foundation for Science, Technology and Ecology, 2000), 85.

5. *Bija—The Seed*, nos. 27 & 28 (2002), 1–2.

6. Vandana Shiva, *Yoked to Death: Globalisation and Corporate Control of Agriculture* (New Delhi: Research Foundation for Science, Technology and Ecology, 2001).

7. Ibid., 2.

8

The Agrarian Mind

Mere Nostalgia or a Practical Necessity?

Wes Jackson

I want to argue not only for the necessity of salvaging what is left of the agrarian mind and way of life, but also for the necessity of its further development and proliferation. When we speak of the need for such a mind, we are not talking about mere nostalgia, but rather a practical necessity. Agrarianism requires no moral or spiritual language for justification; it grows out of a scientific understanding of how organisms interact within natural habitats, an understanding that is too greatly ignored in industrial approaches to agriculture.

The Devastating Assumption of the Modern World

In the last fifty to seventy-five years a disease has spread throughout the world. Those lacking immunity believe that science and technology have the power to solve any problems that science and technology create. Numerous consequences of increasing technical complexity, spurred by scientific discovery, go unmanaged because society can move only on the *available* paths, which are limited in number and design. In other words, the problems generated by the science/technology/economy triumvirate require solutions that would have to evolve along social/political paths that are too small or unavailable.

Agriculture is particularly susceptible. In the April 2002 issue of *The Progressive* Wendell Berry provides a checklist of problems this disease has sponsored: "soil erosion, soil degradation, pollution by toxic chemicals, pollution by animal factory wastes, depletion of aquifers, runaway subsidies, the spread of pests and diseases by the long-distance transportation of food, mad cow disease, indifferent cruelty to animals, the many human sufferings associated with agricultural depression, exploitation of 'cheap' labor, the abuse of migrant workers. And now, after the catastrophes of September 11, the media have begun to notice what critics of industrial capitalism have always known: The corporate food supply is highly vulnerable to acts of biological warfare."

Going through that list, we can acknowledge that soil erosion and soil degradation have been with us since agriculture began some eight to ten thousand years ago. And, yes, exploitation of cheap labor, even slave labor, is an ancient reality. Maybe that is why the temptation to surrender to the forces at work against us is so overwhelming. But what if the social/political conduits are too few and too small to accommodate the much larger pipes delivering the problems derived from the science/technology/economy triangle?

Rather than surrender to industrial society's rampant fanaticism (people in the majority never see themselves as fanatics), how about suggesting that a neo-agrarian mind is a necessity for all? If it has been fair for farmers to be expected to adopt the industrial mind, why is it not fair for urban and suburban folk to adopt an agrarian mind? We have gone one way, why not another?

With this neo-agrarian mind in place, the new developments coming out of science, technology, and economics would be forced to wait or stand in line until the social and political paths have time to evaluate, and if necessary reject, those results or coevolve with them. This neo-agrarian mind of which I speak has two functions: (1) It acknowledges limits most now ignore. (2) It looks to nature as our standard and as a source that offers possibilities we can safely explore.

First, the limits. Limit one: *Soil is as much a nonrenewable resource as oil.* Once destroyed, for all practical purposes in human time, it is destroyed forever. Let your minds travel with me to Mendocino County, California, about four hours north of San Francisco. About every 100,000 years the Pacific tectonic plate slides under the continental plate, creating extensive terraces. There are five such terraces for this story.[1]

Beginning at the ocean's edge, the first terrace is about 100,000 years old, the second 200,000, the third 300,000, the fourth 400,000 and the fifth 500,000. Terrace one features grassland. On terrace two we see lush stands of redwood and Douglas fir. The third terrace is a transition zone of sorts with some Bishop pine coming in. The fourth and fifth terraces will support only what is called a pygmy forest.

By just looking one can see that if one were to weigh up the life forms harvestable on the surface, over an acre say, from terrace two and do the same for terraces four or five, the magnificent redwood and Douglas fir ecosystem of terrace two would weigh much more than the living matter of the pygmy forest. An analysis of the soil will show that the essential land-based elements, those elements necessary to capture the nutrients out of the atmospheric commons, are in a much more limited supply on the older terraces. Without these land-based elements soil cannot be improved. Precipitation over thousands of years has caused a downward tumble—a leaching—of the land-based elements; thus, less life is sponsored.

Why are there not pygmy forests, pygmy prairies, pygmy whatever all over the globe? Well, the earth keeps having these various geologic events, such as mountain formation and the last Ice Age, mostly in geologic time, rarely in human time. These events recharge the surface with elements necessary for life, which once there in combination with life can make soil. Europe and North America had mountain uplifts and the Pleistocene. Uplifts recharge the minerals. The grinding ice of the Pleistocene glaciers pulverized the rock, releasing those essential elements and setting them loose in the biota where nosing roots would capture them to combine with their atmospheric relatives—carbon, hydrogen, oxygen, and nitrogen. We are major beneficiaries of this ice, which came and went over a 1.7-million-year period. Whereas we have been fortunate, Australia has not. Her last geologic uplift was 65 million years ago, and that continent is likely to stay unlucky for a very long time, maybe beyond human time. With such a poor nutrient base, her standing crop of life will never weigh as much as our country's even with comparable precipitation—not until a geologic recharge event. Soil is as much of a nonrenewable resource as oil.

Humans are now the primary earth-moving agents. During the Ice Age, glaciers deposited an average of about 10 billion tons of till in moraines and outwash fans every year. Agriculture today contributes as much displacement as the glaciers of the Pleistocene, and agriculture is not alone. The total movement of earth by humans now is estimated to be around 40–45 billion tons per year.[2] The glaciers gave us fertility by pulverizing the rocks, releasing essential minerals available for life. We agriculturists send those valuable minerals toward a watery grave. Nearly 40 percent of the soils of the world are now seriously degraded. Globally, nearly one-third of the land devoted to farming has been lost to erosion in the last forty years and continues to be lost at a rate of some 25 million acres per year.[3]

The second limit about which the neo-agrarian needs to be informed is related to the origin and evolution of life on our earth. The elements composing living creatures are the elements that are common on the terrestrial surface or within easy reach of roots. In our oceans they are made available through tidal or wave action. Only 18 such elements, which are at once common and necessary for life, exist; the periodic chart lists a total of 105 elements. Carbon, hydrogen, oxygen, and nitrogen are all atmospheric elements that circulate throughout the global commons. The common land-based elements—phosphorous, potassium, iodine, sulphur, calcium, iron, and manganese—are more at home on the back forty than floating in the atmosphere. All of life is the consequence of the complex chemistry arising from these few common elements, 18 out of the 105—less than a fifth. Humans as chemists have used nearly *all* of the 105 elements represented on the periodic chart to create what the advertisers have called "better living through chemistry."

But difference exists between the chemistry of life and the work of human chemists. Life features complex reactants comprising those few elements; chemists feature simple reactants comprising many elements. Professor Terry Collins, director of the Institute for Green Oxidation Chemistry program at Carnegie Mellon University, uses the distinction between nature's chemical world and that of the human chemist as the first consideration in determining what is safe and what is not.[4] He cites as an example an electric eel, made of complex reactants, a complex creature we might eat with impunity. A lead acid battery, by contrast, is a simple creation, but we had better not eat it. While nature creates complexly and safely, we create simply and dangerously.

Professor Collins concludes that the release of human-made compounds into the environment should occur only at rates that are compatible with natural biological processes. Persistent synthetic substances should not be released at all.

If we ignore the implications of this Green Chemistry, we poison the planet and ourselves. The implications for agriculture are profound. At best, 1 percent of applied pesticides reach their intended targets.[5] Since 1950, insecticide usage in the U.S. has increased from 15 million pounds to more than 125 million pounds. Even so, it has been a losing battle: over this same period crops lost to insects nearly doubled from 7 percent of total harvest to 13 percent, and numerous studies have verified the suspected link between pesticides and diseases.[6] Proving a connection has been no small task because direct links are often impossible to establish and experimentation employing direct dosages is usually regarded as unethical; consequently, we are left with epidemiological evidence. A summary of cancer risks among farmers states that "significant excesses occurred for Hodgkin's disease, multiple myeloma, leukemia, skin melanomas, and cancers of the lip, stomach, and prostate" due to pest control chemicals.[7] Another study posits that the herbicide 2,4–D has been associated with a twofold to eightfold increase in non-Hodgkin's lymphoma in agricultural regions.[8] The study of farm chemicals and their evident role in disrupting the human endocrine system is a quickly growing field. One report reveals that several pesticides can reduce the immune system's ability to deal with infectious agents.[9] Formerly, acute poisonings and cancer risks dominated risk assessment. No more. Now, direct evidence from clinical and epidemiological studies shows that people exposed to pesticides experience alteration of their immune system structure and function.

The neo-agrarian will take seriously the implications of Green Chemistry, just as he or she will believe that soil is as much of a nonrenewable resource as oil. Human intervention will be more informed by nature's creative patterns.

A third reality, tiresome to hear once again, is the limitation of the supply of nonrenewable energy, particularly the liquid fossil fuels. This needs to be repeated because essentially all of the natural soil fertility lost to erosion has been more than offset with fossil fuels. The increases in yield during the so-called Green Revolution are fossil fuel dependent. Conventional agriculture is startlingly *inefficient* in energy use, and the trend in countries worldwide is toward even greater consumption of fossil fuels by agriculture to offset the decline in natural fertility. These are brittle agricultural economies that some industrialized societies are putting in place, and the breaking point could be sooner than most suspect. Most energy scholars now project that global oil production will peak and begin its permanent decline between 2008 and 2020, dropping to ten percent of the present annual production by the latter half of this century.[10] The incentive to have more nuclear power plants will rise, and there will be accidents.

It seems that we need constant reminders of the danger of radioactive materials, probably because most exposures don't cause one to drop dead right now. The Center for Disease Control and Prevention (CDC) in the U.S. recently estimated that between 10,000 and 200,000 incidents of thyroid cancer and some 11,000 deaths can be linked to the radiation exposure and potential health effects of tests conducted in the United States and elsewhere between 1951 and 1962. "Any person living in the contiguous United States since 1951 has been exposed to radioactive fallout, and all organs and tissues of the body have received some radiation exposure."[11] This is from testing forty to fifty years ago, and more than health is on the line. The cost to clean up and deal with a Chernobyl-type accident is estimated at more than $350 billion.

Under Murphy's Law and assuming that the probability of a major accident for every reactor is 1 in 10,000, if we had 1,000 nuclear plants we could expect an accident every ten years on the average. Worldwide, we now have 430 reactors, which means an accident about every twenty-two years. Well? We're on schedule.[12]

Who will be the husbands of the soil if the "jig is up" for agrarians? Soil will still be a nonrenewable resource like oil. Chemicals with which we have not evolved must be regarded as guilty until proven innocent. Without natural gas as the feedstock for the Haber-Bosch process, which turns atmospheric nitrogen into ammonia, agricultural production will decline if there is no natural substitute. Soil husbandry thus becomes the primary discipline. Who will be the husbands of the natural fertility of the soil? It will have to be the neo-agrarians backed by their city cousins holding a similar worldview.

The Old Battles

In a real sense, the problems in our time are old problems, but with their modern twist. The aggregation of power is probably as old as humanity. Plutarch, in describing the battle to keep the land in the hand of the small farmers, wrote:

> When the wealthy man began to offer larger rents, and drive the poorer people out, it was enacted by law that no person whatever should enjoy more than five hundred acres of ground. This for some time checked the avarice of the richer, and was the greatest assistance to the poorer people, who retained under it their respective proportions of ground. But the rich contrived to get these lands again into their possession under other people's names, and at last did not hesitate to claim most of them publicly in their own. The poor, who were thus deprived of their farms, were no longer either ready, as they had formerly been, to serve in war, or to be careful in the education of their children; inasmuch that in a short time there were comparatively few freemen remaining in all Italy, which swarmed with workhouses full of foreign-born slaves. These the rich men employed in cultivating the ground from which they had dispossessed the citizens.[13]

The exhaustion of metals and the failure of springs also is an old story. In Roman times, St. Cyprian, Bishop of Carthage around A.D. 250, wrote to Demetrianus, the Roman proconsul of Africa: "The metals are nearly exhausted; the husbandman is failing in his field. . . . springs which once gushed forth liberally, now barely give a trickle of water." Plagues and famines were more frequent, more intense. Demetrianus blamed the Christians for their neglect of the pagan gods. St. Cyprian accused the pagans. Their eroded landscape was never cited as the cause.[14]

Centralized production in cities likewise is old. In 1889 Henry Grady, an essayist and the editor of an Atlanta, Georgia, newspaper wrote up his account of a funeral he had attended eighty miles north of Atlanta. Mr. Grady said that

> The grave was dug through solid marble, but the marble headstone came from Vermont. It was in a pine wilderness but the pine coffin came from Cincinnati. An iron mountain over-shadowed it but the coffin nails and the screws and the shovel came from Pittsburgh. With hard wood and metal abounding, the corpse was hauled on a wagon from South Bend, Indiana. A hickory grove grew nearby, but the pick and shovel handles came from New York. The cotton shirt on the dead man came from Cincinnati, the coat and breeches from Chicago, the shoes from Boston; the folded hands were encased in white gloves from New York, and round the poor neck, which had worn all its living days the bondage of lost opportunity, was twisted a cheap cravat from Philadelphia. That country, so rich in undeveloped resources, furnished nothing for the funeral except the corpse and the hole in the ground and would probably have imported both of those if it could have done so. And as the poor fellow was lowered to rest, on coffin bands from Lowell, he carried nothing into the next world as a reminder of his home in this, save the halted blood in his veins, the chilled marrow in his bones, and the echo of the dull clods that fell on his coffin lid.[15]

This failure to use local resources is not the product of centralized planning so much as of the refusal to expand the boundary of consideration to include the nonrenewable energy cost for production, marketing, and transportation, and the social cost for exporting dollars. The dollar exported will likely not be circulated in the local community or region again. If contemporary sunlight is our source, then local production is the feature. The small quantity of energy captured through the process of photosynthesis in plants is a local source, and to use such a precious quantity to carry heavy loads over long distances is to waste energy on transportation that could be used for food or other purposes locally.

Our present globalization has its analogue in Roman times. According to Ammianus Marcellinus (A.D. 330–395), "for Gallia and the Rhineland towns, it was usual to rely on the harvests of British corn," meaning wheat. In other words, France, Holland, Belgium, and the lower Rhine received this Roman-sponsored grain from England.[16]

THE NEW POSSIBILITIES

To break the hold of these ancient grips and the new problems fostered by the industrial revolution, nature is the only source of possibilities that we can safely explore. Her ecosystems with all of her life forms are products of energy wars going back billions of years, probably even before cellular life. Nature's efficiencies have evolved under the constraints of the laws of thermodynamics. At the ecosystem level, nature features material recycling and runs on contemporary sunlight. She knows how to cope with uncertainty; her tolerance is broad, her strategies mixed, her response rapid when necessary. At many levels from molecules to cells to tissues to organs to organisms to ecosystems, complexity is the order of the day. In a world of increasing uncertainty there is no better time than now to turn to nature in order to learn how she copes with uncertainty.

Wendell Berry speaks of a creaturely relationship with nature. He relates a story told him by his neighbor: The time the neighbor's father was the most tired was when he carried fifty rabbits and a possum up the hill to sell at the small town of Port Royal, Kentucky. When asked why his father hadn't used a horse, the neighbor replied: "Well, we only had two horses and we tried to spare them every way we could." His answer is telling. Contemporary sunlight flowing through muscle tissue united the man and the beast. Empathy ruled.

A creaturely life, using contemporary sunlight, has been a source of culture. More than three decades ago, I heard a lecture by the late Georg Borgstrom, a well-known nutritionist. Speaking of the origin of cheesemaking in Europe, he said that in a sun-powered culture, using the muscle power of draft animals and humans, one doesn't haul all of that water to town. It takes about ten pounds of water to yield one pound of cheese. Saving the load factor on the legs of an ox or a horse by a factor of ten is, or course, not trivial.

With empathy as a product of a creaturely life, followed by cultural accouterments, a creaturely world also features resilience. December 2001 ended the last field/pasture season for the ten-year experiment called the Sunshine Farm at The Land Institute in Salina, Kansas. Marty Bender directed this project, which involved 210 acres: Fifty acres of our area's conventional crops were grown in a five-year rotation in the most ecologically correct manner we could muster. A quarter section or 160 acres was pasture devoted to our longhorn cattle herd. One purpose was to determine what the energy cost would be if we were to use the equivalent of the contemporary sunlight falling on that acreage to sponsor all farm activity. A photovoltaic array met the electrical demands of the farm, and two oilseed crops—soybeans and sunflowers—grown in the rotation were the energy sources for the biodiesel tractor. The first five years we also had draft animals, and so Marty Bender was able to compare the energy needs of the tractor to those of the draft horse. The embodied energy of every purchase was estimated, whether it was for mining the ore in the Minnesota iron range to make the tractor or to make a one-pound bolt. Preliminary data suggest that roughly 25 percent of the till acreage must be allocated for traction and 40 percent for sponsoring mostly nitrogen fertility.[17] Though the data are not completely analyzed at this writing, it does appear that the requirements of crop acreage for the horse and the tractor may not be too dissimilar. One can shut off the tractor and not have to pay it to stand around and be a tractor. On the other hand, a pregnant mare at work can be growing her replacement. One does not come out some morning and find a little baby tractor. Moreover, during a drought a tractor will starve faster than the draft animal. The tractor "grazes" on a narrower photosynthetic base using high-quality oil and carbohydrate-producing crops. The draft animal has an enzyme system, making it more of a generalizer that tolerates, indeed thrives, on varying conditions; it is more resilient.

An Agriculture for the Neo-Agrarian

The science in support of the neo-agrarian has two components. Ecologists and evolutionary biologists have wanted to understand how the world *is*. Agriculturists have needed to understand how the world *works*. The cognitive difference between *is* and *works* is so subtle that the two are usually interchangeable.

For agriculturists, how the world *works* must be of primary interest. Utilitarianism necessarily rules even though it may limit our imagination. The utilitarian side is under the control of a dominant social structure. How the world *is* is part of the human experiment or journey to answer the old religious questions, such as Where did we come from? What kind of thing are we? And what is to become of us? The nature of materials and energy, gravity, and life processes are all part of this understanding. Ecology and evolutionary biology are among the disciplines derived from the "how the world *is*" motivation.

The nature/technology (*is/works*) dualism sponsored by Enlightenment thinking must and can come to an end. We see the possibility as we research a new agriculture at The Land Institute, where we look to nature's ecosystems because they feature material recycling and run on contemporary sunlight. Because industrial agriculture depends on fossil fuel subsidies for fertility and traction, minimizes material recycling, and leads to chemical contamination of our land and water, if we are to look to renewable energy to sustain us, nature's ecosystems are the most relevant standards.

Our major effort at the moment is perennializing the major crops. Wheat, rye, and sorghum are the leading species, followed by sunflowers. Once perennialized, these domestic species will be arranged as domestic analogs of the four groups common everywhere we find prairie: warm and cool season grasses, legumes, and members of the sunflower family. We are also domesticating Illinois bundleflower, a wild, perennial, high-yielding legume. We will one day plant out the first domestic perennial prairie featuring such major crops as wheat, rye, sorghum, and sunflowers (and eventually other crops) as part of a diverse system that will function like a prairie. These mixes of perennial roots will hold and build soil. By featuring species diversity, we achieve chemical diversity requiring a tremendous enzyme system on the part of an insect or pathogen to cause an epidemic. By putting legumes in the mix, we get biological nitrogen fixation.

We are confident that in twenty-five to fifty years this Natural Systems Agriculture can make significant strides toward sustainability through the merger of ecology and evolutionary biology with agriculture. Nature's economy can then one day begin to inform the human economy by means of an agriculture that operates in ways analogous to a native prairie or forest.

As this new agriculture begins to operate, dislocations of the current economic order are certain. Here will be a system in which the reward runs to the farmer and the landscape, rather than the supplier of inputs. The suppliers of commercial fertilizer, currently dependent on natural gas as the feedstock to make ammonia, will have greatly reduced sales. The chemical industry marketing pest control substances will decline because of species diversity. The seed houses will experience reduced sales with perennials in the field. With reduced plowing and tillage the market share for farm machinery companies will decline. These economic problems will prove to be more manageable than the large-scale bookkeeping fraud witnessed recently. They will at least put our economic priorities in line with ecological necessities.

The neo-agrarians who manage this sustainable food economy will be growing the prototype for a different economic order, one that will feature material recycling and the use of contemporary sunlight. Available across the ecological mosaic of our agricultural landscape, this model will highlight the necessity of minimizing the number of loops we widen away from the farm. Local adaptation will become agriculture's prevalent feature for both ecological and food reasons. The prideful notions inherent in globalism will wilt and die.

By implementing such a worldview, we will have drawn our boundaries of consideration to more completely overlap the boundaries of causation.[18] The cost of cancer treatment will be included in the cost for the use of pesticides, causing their use to plummet. The health of soil, air, water, and humans will be one subject. Such an ecological-evolutionary view implies the following: "Don't feed dead sheep to cattle" and "Don't use antibiotics for growth promotion." Without large feedlots, the massive doses of antibiotics will become unnecessary and genetic resistance of the microbes will cease. With no major feedlots, cruelty to animals will decline and so will feedlot pollution.

Finally, such a worldview will make us appreciate the value of keeping the cultural seed stock—rural people—in place, on the farms and in the small towns and rural communities. The language of economic determinism, which has encouraged the exodus from the farm and village, will no longer have currency.

The questions we must ask, then, are these: Is neo-agrarianism ecologically and culturally necessary? Are we not dealing with a practical necessity? After all, the loss of agricultural land and the loss of agrarian culture amount to the loss of our future food supply.

NOTES

1. Jenny, H. 1980. *The soil resource*. New York: Springer-Verlag.

2. Hooke, Roger LeB. 1994. On the efficacy of humans as geomorphic agents. *GSA Today* 4 (9):217, 224–25.

3. Currently more than 2 billion hectares of soil have been degraded worldwide by human activities. See especially, Wood, S., K. Sebastian, and S.J. Scherr. 2000. Pilot analysis of global ecosystems: Agroecosystems. Washington, D.C.: IFPRI and World Resources Institute. Also see UNEP. 2002. GEO-3: Past, present and future perspectives. London: Earthscan.

The percentage of degraded land, based on worldwide estimates, ranges from 15 to 38 percent of cropland degraded by agricultural activities over the last fifty years. See Scherr, S.J., and S. Yadav. 1996. Land degradation in the developing world: implications for food, agriculture, and the environment to 2020. Washington, D.C.: IFPRI. Also see Wood, Sebastian, and Scherr 2000 and UNEP 2002.

4. Collins, T. 2001. Toward sustainable chemistry. *Science* 291 (5501):48–49

5. Pimentel, D. ed. 1991. *Handbook of pest management in agriculture. Vols. I-III.* 2d ed. Boca Raton, Fla: CRC Press.

6. Bell, E.M., I. Hertz-Picciotto, and J.J. Beaumont. 2001. A case-control study of pesticides and fetal death due to congenital abnormalities. *Epidemiology*

12:148–56. Hardel, L., and M. Eriksson. 1999. A case-control study of non-Hodgkin lymphoma and exposure to pesticides. *Cancer* 85:1353–60. Lu, C. et al. 2000. Pesticide exposure of children in an agricultural community: Evidence of household proximity to farmland and take-home exposure pathways. *Environmental Research* 84:290–302.

7. Blair, A. et al. 1992. Clues to cancer etiology from studies of farmers. *Scand. J. Work Environ. Health* 18:209–15.

8. Blair, A., and S.H. Zahr. 1991. Cancer among farmers. *Occupational Medicine* 6 (3): 335

9. Repetto, R., and S. Baliga. 1996. *Pesticides and the immune system: The public health risks.* Washington, D.C.: World Resource Institute.

10. Hatfield, J.L., and D.L. Karlen. eds. 1994. *Sustainable agriculture systems.* Ann Arbor, Mich.: Lewis Publishers; Hall, C.A.S., C.J. Cleveland, and R. Kaufmann. 1986. *Energy and resource quality: The ecology of the economic process.* New York: Wiley Interscience.

11. News in brief. 7 March 2002. *Nature* 416:8.

12. Barkenbus, J.N., and C. Forsberg. 1995. Internationalizing nuclear safety: The pursuit of collective responsibility. *Annu. Rev. Energy Environ.* 20:179–212.

13. Jacobs, H.E. 1997. *Six thousand years of bread.* New York: Lyons Press.

14. Ibid.

15. Jacobs, J. 1993. *The death and life of great American cities.* 2d ed. New York: Modern Library.

16. *Six thousand years of bread.* Jacobs says the grain ships brought the grain from Sussex and Kent.

17. Bender, M.H. 2003. Energy in agriculture: Lessons from the Sunshine Farm Project. In *Proceedings of the Third Biennial International Workshop, Advances in Energy Studies: Reconsidering the Importance of Energy, Porto Venere, Italy, 24-28 September 2002,* edited by S. Ulgiati. Padova, Italy: Servizi Grafici Editoriali. In press.

18. Levins, R., and R. Lewontin. 1987. *Dialectical biologist.* Cambridge, Mass.: Harvard University Press.

9

All Flesh Is Grass

A Hopeful Look at the Future of Agrarianism

Gene Logsdon

When I try to define or at least describe agrarianism in a way that is useful to me, I think of bib overalls. Bibs are an invention of agrarianism, and like agrarianism, refuse to go away despite all fashion wisdom to the contrary. Bibs, the uniform of the farmer behind his team of horses a hundred years ago, remain the most frequently worn uniform of the farmer today in the cab of his three hundred horsepower tractor. More surprising, bibs have, since the 1960s, crossed that supposedly impenetrable cultural barrier between farm and city to become and remain a popular style of clothing among urbanites. I have no doubt that bibs will be around a hundred years from now, mainly because they are comfortable and useful. They may be see-through bibs, but you will still be able to carry half a hardware store around in them.

Agrarianism will still be around too, though it may look quite different from what it looks like now. The forces arrayed against it are mighty, as are the forces arrayed against bib overalls. It is easy to predict more monopolies taking over the food business today, easy to point out that a few huge mega-companies already control the food supply, easy to see farmland falling inevitably into the hands of a new landed oligarchy. But it is also easy to realize that this drift toward the total consolidation of power will collapse, because historically it has always collapsed. We are following unerringly in the footsteps of the old Roman Empire.

To have a hunch, as I do, that most of the food we actually eat in the future will come from gardens is no more ludicrous than believing, as agribusiness does, that most of the food will come from the likes of Cargill and DuPont, with their lunatic slogans of "from seed to shelf" and "from dirt to dining table." Millions of people in Asia live from the production of farms no bigger than large gardens. And what more sustainable or stable food production system could be envisioned than a landscape of garden farms from inner city to the outermost countryside? It may be true that a mechanic building his own car in his garage can't compete with Ford or Honda, but no Cargill or DuPont can raise a squash cheaper than a backyard gardener. Russia learned that; so shall we.

There is something else going on today that, in conjunction with a nation or world of garden farmers, is bringing on a flowering of an old-new agrarian way of producing food. Like the garden farms of Asia, this old-new way of farming is not pie-in-the-sky dreaming, but an established, proven method. I am referring to pastoral farming, where grazing animals provide meat, milk, eggs, dairy products, wool, and hundreds of other animal products at a fraction of the cost of producing them with current factory technology. It is being called "grass farming," or "managed, intensive grazing," or the term I've been using, "pasture farming."

Pasture farming is the first alternative to high-tech agriculture that has both short-term and long-term profit on its side. Industrial grain farming and animal factories may have had short-term profit advantage for awhile. That's why they rose to prominence. But no longer. Over half the money to keep these operations afloat now comes from government subsidies. If there is any lesson of history that always remains true, it is that no economy can be falsely sustained when it can't compete with another economy. Neither all the king's horses nor all the king's men can make industrial farming survive with subsidies, just as fifty years of subsidies could not save the old family farm. Industrial farming is simply not profitable enough anymore to compete with pasture farming.

It takes only a cursory comparison of today's industrial farming with pastoral farming to show how the latter's day has come again. In a typical industrial grain situation, the farmer goes to the fields in spring with enormously expensive machines to plow or disk or chisel the soil into a seedbed and/or to spray it with herbicides to kill the weeds. However arrogantly this modern farmer prances his huge machines across the pages of his agribusiness magazines, he is at the mercy of the weather more than any farmer in any previous era. If rain delays planting beyond a rather short window of opportunity, the corn and soybeans will not make the high yields that the farmer must get if he is to make a profit. He sows expensive bioengineered seed frantically between spring showers with huge, expensive planters and applies expensive fertilizers that are made with natural gas or hauled from mines hundreds, even thousands, of miles away. Then more spraying. More praying too that rains do come now that the crop is up, but not too much or too little, which would again rule out the high yields he needs to make a profit. Hopefully the rains are not accompanied by hail or floods to destroy the growing crop. To cover himself, the farmer buys expensive insurance. At harvest time, into the fields he goes again with huge, enormously expensive harvesters, praying for good weather, not mud that will compact at great loss of productivity under the weight of his dinosaur machines. The grain goes from harvester into expensive semi-trucks and is hauled to elevators or to on-farm storage, where it must be binned, dried (with natural gas), then watched so grain-destroying insects don't infest it, all at high cost. Then the grain must be loaded and moved again, by rail or truck or barge, at a great consumption of fuel and other transportation costs, to very expensive animal factories where it is ground into meal in very expensive mills and fed to animals that are kept healthy in their crowded quarters by costly (in more ways than one) hormones and antibiotics. The manure, sometimes in quantities as large as the sewage output of large cities, must be somehow handled, stored, and finally hauled out and applied on soil or gotten rid of some other way, all to the constant protest and nuisance suits of neighbors angry about the odors and flies. Then much of the meat, milk, and eggs must be shipped back to where the grain came from.

Compare that with raising the same meat or eggs or dairy products in a pasture farming system. Assuming that the fields of grasses and clovers and other grazing crops have been established, spring work amounts to turning the animals onto the pasture paddocks in rotation and watching them eat. Rains do not hamper soil cultivation because there is no soil cultivation. There is no erosion. There are no costly cultivation tools. Hail will not hurt the grass. Even flooding only harms it temporarily. The animals harvest the pasture crops, control most weeds by eating them, and spread their manure for fertilizer, all without labor, fuel, or machine expense. The amount of fertilizer and herbicides necessary to keep the pastures productive is minimal, sometimes not necessary at all. The farmer, or grazier, the preferred term, mends fences and makes hay. If he gets good at grass farming and employs all the pasture plants now available, he will have to make only a minimal amount of hay, just enough to supplement grazing in winter. When he gets really good at grass management, the animals can pasture year round except after heavy snows or, in some soils, for a short time during spring thaw.

I don't think you have to be a genius to figure out which farming method is more economical, as well as ecologically sane.

Had not the increasing years forced me to use my head instead of my back, I might not have discovered that there are easier and cheaper ways to be a "tiller of the soil" than to actually till the soil. All I had in mind was a practicality: I wanted to continue to pursue farming experiments even into the days of failing strength. What I needed was a way to get the work done and the food produced without so much time, sweat, or potential hemorrhoids from riding tractors all day long. Reading Masanobu Fukuoka's classic, *One Straw Revolution*, I realized I was not alone. Wrote Fukuoka: "I was heading . . . toward a do-nothing agricultural method. The usual way to go about developing a method . . . results in making a farmer busier. My way was opposite. I was aiming at a pleasant, natural way of farming which results in making work easier instead of harder . . . I ultimately reached the conclusion that there was no need to plow, no need to apply fertilizer, no need to make compost, no need to use insecticide. When you get right down to it, there are few agricultural practices that are really necessary."[1]

Fukuoka was not by far the first farmer to challenge the myopic work ethic that wealthy people urge upon society, the better to ensure more wealth for themselves. In 1949 Channing Cope, in a book he called

Front Porch Farmer, told how he had achieved year-round grazing on his seven hundred acres in Georgia, allowing him to do most of his farming from his porch swing while watching the livestock do the work. No more of what he called "drudge farming" for him. To prove that Cope was not just being funny, many other books came out in the early 1950s that proved as conclusively as anything can be proven about agricultural practices that grassland farming, as it was called then, returned more profit per acre than grain farming, not even counting the environmental profits. The last word on the subject came in a seven-hundred-page tome edited by H.D. Hughes et al. at Iowa State University in 1951 called *Forages*.[2] Some fifty experts in the book attested to the proposition that a farmer could raise meat, milk, butter, cheese, wool, eggs, and other animal products with less work and expense by pasture farming than by a production system based on annual tillage of grain crops.

Not that any of these books plowed new ground. As early as 1775 a farmer in Scotland, James Anderson, described in his journal (farmers actually kept journals then, not just account books) a practice of permanent pasture farming that was almost exactly like the strip-grazing methods of grass-farming dairymen today: "A farmer who has any extent of pasture ground, should have it divided into fifteen or twenty divisions . . . and the beasts be given a fresh park each morning, so that the same delicious fresh repast might be repeated daily." This kind of farming, Mr. Anderson insisted, would bring more profit than Jethro Tull's misconceived notions of annual cultivation of row crops, which was then becoming more common.

But Jethro Tull won the day and so did corn farming in the 1950s, even though the premises upon which both were based are wrong. Tull's error was ludicrous. He argued that plant roots had tiny sucking mouths to consume microscopic soil particles, and the more cultivation broke up these particles and pressed down into the plant root mouths, the better the plant would grow. Cultivation of ungrazed grains was of course necessary for weed suppression, but Tull and three hundred years of plowmen after him then made the wrong deduction: that pulverizing the soil itself is the reason plants grow better, rather than the weed suppression that follows. Nor did the row croppers ever understand that in a grazing regime weeds were not nearly as much of a problem because the animals, by eating them, kept them from suffocating the grass and clovers. In fact, in many instances, the weeds are as nutritious as the grasses and clovers. In the 1950s, more for reasons of consolidating power, the government started subsidizing cultivated grain production but not grazing, and so the corn monopoly prevailed. The government has rarely subsidized grazing because graziers do not depend on agribusiness like grain farmers do. Government subsidies are meant to help agribusiness, not foster an independent, efficient farming culture.

I asked Wes Jackson, a geneticist and farmer who is developing his own version of a pasture system at his Land Institute in Salina, Kansas, why his work does not gain more attention from commercial agriculture. "From the moment humans first touched plow to soil, exchanging hunting and gathering for domestic agriculture, we committed ourselves to ultimate decline," he said to me (and wrote in his *Altars of Unhewn Stone* in 1987). "It is a tragedy in Alfred North Whitehead's sense of tragedy, the remorselessly inevitable working of things. Given the current human population now dependent on till agriculture, we will need to continue to till the earth, even though such activity has historically and prehistorically undercut the very basis of our existence."

But as Dr. Jackson knows and bets his career on every day, the situation may not remain "remorselessly inevitable." Farmers all know now that tearing up the soil and tilling it is not necessary for good crop growth. No one tills the forests to grow trees. No one tills the prairies to grow grass. Plants mature, drop their seeds on the soil surface to sprout, and grow again. Tillage may be necessary to control weeds but not to "fit the soil." In fact, tillage over the years is what makes the soil unfit. Herbicides (for better or for worse) prove that point. Farmers can grow excellent crops with what they call "no-till" farming, using herbicides rather than tillage to control weeds. But so far only a few farmers have made the corollary deduction: in pastures, weeds seldom compete critically with grass and legume crops as they do in cultivated grain crops, and they can be mostly controlled by grazing animals and an occasional mowing for hay.

The notion that annual tillage of grain is agriculturally necessary to keep the world's cupboards full of bread, meat, and dairy products led to another myth—that animals need to eat cultivated grains to make them grow or produce "efficiently." The farm animal market developed on the basis of these two misconceptions, and then institutionalized them. Today we live under a corn and soybean monopoly. Prices for meat are reckoned on a grading system that values the meat by the amount of corn and soybean meal stuffed into the animal. My baby beef, consuming only mother's milk and good pasture for seven months and then butchered, is just as tasty as any corn-fed steak in Omaha. But it won't bring Choice or Prime prices, not because of taste, as the consumer is lead to believe, but because the meat packers can butcher and process one 1500-pound steer faster and at less labor than it can process two 650-pound baby beeves. Until that situation changes, succulent, healthful milk- and grass-fattened meat is not going to dent the corn monopoly marketplace.

Critics worry that pasture farming can't produce enough food to "feed the world." In fact quite the opposite is true. There are millions of acres in the world that are amenable to grazing, but are not cultivatable for annual grains. If we posit an ever-increasing population, as the "feed the world" proponents do, pasture farming this land is an absolute necessity.

That should be the end of the argument, but let us pursue it a little further. Many other acres now devoted to annual grains do not produce as much food as they would if they were converted to managed grazing lands. The literature of grass farming is full of examples documenting this fact. Moreover, as grassland these acres can be *sustained* in food production, whereas, if doggedly cultivated every year they will inevitably decline in fertility because of erosion or other types of resource depletion. Who will these acres feed then? Groundhogs? What will happen to the irrigated corn acreage of the Plains when the wells run dry? What will happen in the eastern half of the nation, where there is plenty of water and government subsidy to encourage unprofitable grain production, if farmers doggedly keep on planting annual grains on erosive hills that in dry years produce meager crops, or plant on flood-prone bottom lands that in wet years produce nothing?

Only on the very best farmland, of which there is a fairly limited amount, can annual grain out-produce grazing, and even that is doubtful when everything is considered. Dairy cows may produce more milk on a heavy grain diet (because they are bred that way) than on grazed forages, but not necessarily. What generally happens to counteract this possibility is that the grazier will keep a few more cows to make up the difference, which in a grazing regimen is much more economical to do than trying to boost production per cow. The production per acre is just as high, and often higher, on the grazing farm even if the production per cow is lower. In terms of "feeding the world," what's the difference?

The whole issue of "feeding the world" seems specious to me. What does it mean, actually? All my life in farming I have heard government urge us to "gear up" to "feed the world." It sounds so noble and we fall for it because we think it means we will finally make some money. Now I understand that the expression is merely a euphemism for "Push American grain overseas and keep grain cheap here so the American consumer can afford to buy more cars, television sets, houses, and ten trillion gadgets." We "geared up," we raised bumper, surplus crops and still people all over the world starved to death. Today "feed the world" is the forked-tongue hypocrisy that mega-companies utter while they try to monopolize the food business.

No country, no company, no government can feed the world, especially when the cost of the food is greater than the people who need feeding can afford. Food is a much more complex thing than the "feed the world" enthusiasts want to understand. Here are a few random thoughts that ought to make them wonder a little: (1) Much of the world lives quite well without corn and soybeans or animal products fed with corn and soybeans. American grain farmers couldn't "feed" them if they wanted to. (2) Millions of people eat foods Americans have never heard of. The Mongolians thrive on horse milk. (3) I've a notion that there is enough land worldwide in lawns, golf courses, roadsides, and parks to "feed the world" if the most advanced raised-bed gardening methods were used. (4) I know for a fact that our family could easily supply all our meat needs from the wild animals that live on our little farm. Our neighbor, a wild food enthusiast, says there are so many wild edible and nutritious plants growing in fencerows and fields and parks and vacant lots and roadsides that no one should ever starve to death. (5) Summing up: "Feeding the world" is not the problem. Teaching people how to feed themselves is the problem. But agribusiness doesn't want people to know how to feed themselves.

The bias in favor of corn-fed is particularly held sacred by hog producers, who like to point out that hogs have only one stomach and so unlike ruminants they can't utilize grass "efficiently." This contention is not really true beyond the moot point that hogs can get fat faster on an exclusive diet of grain and soybean meal. A lot of hogs are still raised at least partially on pasture. I once worked for a very successful farmer who liked to turn his pigs at weaning time into alfalfa so tall that the pigs literally disappeared into it. They ate it down to the ground and reached market weight on less than half the corn that they would normally have been fed. Also, there is a vigorous market for fifty- to one-hundred-pound pigs for barbecuing. Such pigs can be raised on mother's milk and pasture and a little grain. And the grain does not have to be row crop corn and soybeans, but pasture-seeded barley or oats, or for that matter, acorns in the woods. But again the meat packers aren't interested in ultra-delicious suckling or barbecue pigs. They scoff at the idea of butchering a small, fifty-pound pig as being inefficient while they accept unhesitatingly butchering chickens that weigh only five pounds!

And of course, range-fed hens can lay as well as grain-fed hens. Broilers can fatten on range too, just not as quickly. The name "fast food" is right on more than one account.

Dairymen believe they can't attain high milk production without feeding cows heavy rations of corn, and yet they admit freely that milk production goes up when cows are turned out on good pasture. Believing in grain, the dairy industry has bred cows that can take advantage of heavy grain feeding. What if agriculture had been breeding cows that could take better advantage of grazing crops? In any case, that whole argument is out the window if pasture farming advances to where the grain can be grown as a graze crop instead of harvested, dried, stored, and ground at enormous expense, as I have demonstrated is possible with wheat and oats. David Zartman, a scientist at Ohio State University, is demonstrating that cows can graze even corn on the stalk, thereby bypassing the entire corn harvesting extravaganza of high-tech agriculture.

Most farmers resist the advantages of pasture farming and stick with erosive, annually tilled crop farming and disease-prone animal factories because that's what they know and that's what made some financial sense before the costs of industrial grain farming skyrocketed. Now they are locked into huge investments in machinery and buildings, often not paid for, that pasture farming would render obsolete. They *can't* change. Also, cash grain farming seems to allow for about three months of relatively little work on the farm (although many such farmers have to maintain off-farm jobs to fund their addiction to corn and soybeans). But they prefer this kind of living, hectic as it is, to being "tied down" by livestock. They are disbelieving when told that pasture husbandry allows for as much free time as cash grain tractoring. Farming from the seat of a huge, air-conditioned tractor appears much more exciting to them than farming from the front porch. "If I had to go back to raising livestock, I'd just quit farming," says my neighbor.

And that is exactly what many cash grain farmers are being forced to do. They are in denial. They can't believe the changing economic situation. Meeting after meeting is being held to find a cure for the economic malaise. The only answer that all agree on, and it is stated in all seriousness, is for nature to deliver a bad crop year. That's how far this insanity has gone.

There seems to be no end to the madness. As grain prices sputter from overproduction, costs continue to rise:

1. Cash grain farming survives on cheap fuel. Cheap fuel can't last.
2. The cost of farm machinery has become obscene. New big tractors and harvesters at two hundred thousand dollars, more or less, are so unaffordable at present grain prices that farmers never get them completely paid for. Used machinery is rolled over into new equipment, and dealers endeavor to make up the difference by selling the roll-overs to less well heeled farmers who still can't afford them unless they roll over their older (and usually unpaid for) equipment to farmers farther down the income ladder. The October 2000 issue of *Farm Journal* magazine quotes Jim Irwin at Case-International Harvester as saying, "It now takes about 1,000 individuals to move the used farm equipment from the top seven customers in North America through

the system, so it's no wonder that used equipment is a problem" (9). Pasture farming gets rid of iron-fever disease.

3. The amount of land good enough to grow cash grain profitably (if "profitable" is any longer a meaningful word in grain farming) declines because competing users can afford to pay more for it. Builders of highways and homes and factories would rather build on level, well-drained, deep soils, not on hillsides or low-lying poorly drained soils. Pasture farming can utilize poorer soils more profitably than grain farming. Between all the competing uses, where will the corn growers go?
4. The corn monopoly is looking to ethanol to save its market as the supply of oil diminishes and/or goes up in price. But corn is not the most efficient crop for ethanol. Sorghum is better. Corn is being pushed because we live under a corn monopoly. More and more, ethanol science is looking for biomass from perennial grasses. The latest news out of the biotech labs is that geneticists have engineered genes in grasses that make distilling ethanol from permanent pastures more efficient than from annual corn.
5. Animal factories, the chief beneficiaries of the grain subsidy program because they can buy grain cheap on the market while the government makes up the difference to grain producers, have racked up a horrendous record of financial and environmental disasters. Attempting to handle manure in what is essentially a flush toilet system, using water to move the stuff instead of bedding, just doesn't work well without a sewage treatment plant. But animal factories are not yet required to install sewage treatment plants or effective composting plants. This is insane, but if treatment plants were so required, animal factories would no longer be able to compete with pasture husbandry at all. In pasture farming, the management of manure is done for free by the animals themselves. Farm factories are now competitive only because the government subsidizes "manure management."
6. Using antibiotics routinely to keep factory animals healthy has finally crashed into the wall of the industry's own shortsightedness. In 1995, poultry factories, desperate for a new antibiotic that disease bacteria were not becoming immune to, begged the FDA to let them use

fluoroquinolones, one of the most valuable antimicrobial drug classes available to treat human infections. Over the protests of many scientists and medical doctors, the FDA acquiesced. But just recently, the FDA withdrew its approval. I quote from its own press release: "The FDA has determined that use of these fluoroquinolones in poultry has caused the development of fluoroquinolone-resistant campylobacter, a pathogen to humans . . . *this fluoroquinolone-resistant compylobacteris transferred to humans and is a significant cause of the development of resistant campylobacter infections in humans*" (emphasis added). Where will the animal factories turn when antibiotics (not to mention equally problematic growth hormones) are no longer effective?

7. The latest research reports say that annual soil tillage is a big contributor to the buildup of greenhouse gases, especially CO_2. Score another big plus for pasture farming.

Eating animal products from a grazing regimen creates major nutritional and health advantages. Alan Nation, who has sturdily led the shift to modern grass farming with his *Stockman GrassFarmer* magazine, writes in the September 2000 issue:

> I shocked a group of graziers the other day in saying that in 20 years all of the grass dairies and grass finishing operations would be on Class I farmlands. The same quality soil that will grow 185 bushels of corn is the same quality soil that will best grow the necessary . . . dairy and beef-finishing forages. New Zealand's famous Waikato Dairy District and Argentina's beef finishing zone [both based on pasture-only farming] are those countries' best land, not their worst. . . . [But] the naysayers ask, "Why go to all that trouble when you can just feed grain for the extra energy?" There are a lot of arguments for pasture only, but today I will use the simplest one. Grain feeding is going to increasingly prevent you from taking advantage of premium-priced market options. Research has found that feeding as little as five pounds of grain per day cuts the CLA [conjugated linoleic acid, which holds great promise in fighting cancer, obesity, and diabetes] content of the milk in half! Also in research, grass finished beeves had

> 250% more CLA in their intra-muscular fat than grain-finished beeves. (10)

Meat and eggs raised on pasture versus heavy grain rations also have less cholesterol, lower levels of acid-resistant *E. coli*, less fat, more essential omega-3 fatty acids, as well as higher levels of various anti-cancer agents, and more essential vitamins. (The hottest book in agriculture right now is Jo Robinson's *Why Grassfed Is Best,* which details and documents these claims.)

Some scientists in the land grant agricultural colleges in the corn belt, which generally resist the notion of a food production system based on pasture rather than cultivated grain (they know where their money comes from), are slowly changing. David Zartman, the champion of pasture farming at Ohio State, believes the industrial grain/animal factory dominance in American agriculture will last only one more generation before moving to Brazil and playing itself out. The University of Missouri economist John Ikerd writes: "This era of cash grain industrialism and giant animal factories could be about over."

Bob Evans, the hard-headed businessman who made a fortune with his fast food restaurants, but who has been before that and since then a cattleman who practices year-round grazing on his farm in southern Ohio, goes around the country preaching grass farming as the way to save the farm economy. "You don't need those big tractors and harvesters, don't need all those fertilizers and chemicals, don't need all that grain, don't even need much hay," he says, and he has demonstrated on his farm the truth of his claims.

To convince myself that till-less pasture farming is not just a pipe dream, I began visiting farms that practice it. David Miller, an Amish farmer in Holmes County, Ohio, convinced me, if Bob Evans hadn't, that at least pasture beef production is supremely practical. His whole farm of about three hundred acres is a permanent pasture and beautiful beyond words. The place looks like a golf course with a couple hundred head of cattle wandering over it instead of golfers. The cattle are sleek, in such good condition that it seemed to me, brainwashed by growing up in the corn belt, that corn had to be a big part of their diet. Miller smiled. "I don't grow any. I buy a couple of tons for winter feed." I was doing rapid calculations in my mind. "But," I said, "that would mean less than a bushel of corn per cow, not enough to make any difference." "Yeah, that's about right. I don't think I'd have to feed any corn. I give them a little, sort of, well, for a treat, like candy for the kids."

I visited Bruce and Lisa Rickard, fugitives from the computer world and so unbiased by the corn monopoly mind-set. They make a living with sheep and cattle on a farm entirely given to pastures. There's not a stalk of corn on the place. They graze rotated pastures year around with supplemental hay as needed in winter. They own only enough machinery to make hay and figure eventually they will not need to do that either.

Nathan Weaver, a dairyman in Holmes County, showed me how he uses managed, intensive grazing to save money and increase profits from milking cows. (He writes on how to become a pasture farmer regularly for *Farming* magazine of Mt. Hope, Ohio.)

I went home profoundly impressed. The shepherd and the cowboy had it right in the first place. I changed our little place into a pasture farm. It was mostly that anyway. I eventually divided about fifteen acres into ten paddocks, five of about two acres and five of about one acre, so I could control the grazing by rotating the livestock from one paddock to another. In 2000 and 2001 I grazed sheep all of November and December and part of January, and a little even in February and March. This was unheard of in this climate in my father's day. And I have hardly begun to learn and implement the science of year-round grazing. To watch what is happening, and to ponder what it could mean if agricultural science had been improving crops for grazing as assiduously as it has been improving corn for mechanical harvest, sends my pulse to racing.

Every day I see more possibilities. I found that at the end of the second year of red clover, when the plants decline, there is a time when the ground is relatively bare. If I broadcast wheat (barley or rye or oats work, too) in the declining stand in September with a little broadcast seeder, it sprouts, roots down, and gives me a stand that I can graze or harvest for grain, no cultivation involved at all.

I then learned that I could make a crop of hay from a stand of oats before it went to head and, when the oats regrew and headed out, pasture it as a grain supplement. And then, to my utter amazement, a third crop emerged from oat grains that the sheep missed. This third crop provided green pasture even into early January. I began to understand why gatherings of graziers are so full of excitement compared to the sullen visages of farmers at grain conferences. We know that we are on to something revolutionary and hopeful.

The environmental benefits of pasture farming over annual cropping are far-reaching and perhaps incalculable. Not the least benefit is that pasture farming results in a much more delightful and pleasant environment in which to live. There are no awful odors and fly infestations or roarings of huge machinery that plague rural areas (and drive down the value of rural properties) wherever industrial farming prevails. As I walk across a pasture of sweet-smelling ladino clover, I'm aware not only of many species of songbirds around me (the bluebirds hollow out nests in the wooden fenceposts) but of the soothing hum that hangs over the blossoms as my bees work hard to supply us with honey, my favorite dessert, for free. Their pollination of the ladino makes it seed enough so that even with pasturing, enough seed falls to the ground to keep the ladino growing continuously without any work from me. I look across the fields and enjoy the roll and spill of grassy hillside meadows where once there were gullies and thank the permanent pastures that our creek remains spring-fed while most of the springs where intensive grain farming persists have dried up long ago.

The fish in our pond are also a part of our grazing regimen. They graze the water as the livestock graze the land. The water runs off the grass clear, not muddy like the erosive waters off the nearby cornfields. Our fish are part of our food supply. All we have to do is catch them, a form of recreation, not work. We don't even feed them a supplemental grain-based fish food. They won't eat commercial feed because the pond, which we manage as carefully as the pastures, provides them a self-sufficient supply of natural food.

Agribusiness friends tease me about wanting to "go back" to a pastoral, hunting and gathering stage of civilization because of my interest in a front porch agriculture. That may be an accurate assessment except that I would say "go forward." History and anthropology indicate that domestic farming and hunting and gathering are seldom separate "stages" of civilization as we are sometimes led to believe in our schooling. Most previous societies actually practiced a combination of hunting and gathering and domestic agriculture for extended periods of time. There is evidence that the "golden eras" of civilizations occur under such circumstances. But civilizations don't maintain the balance between hunting and gathering and domestic agriculture that results in a pastoral or agrarian society. They move on to a totally cultivated and centralized agriculture and get fat. A leaner, meaner civilization then conquers them. So civilizations rise and fall, cycle round and round, the same old story century after century.

Jared Diamond, a physiologist and evolutionary biologist at UCLA has most recently addressed this subject in his intriguing book, *Guns, Germs, and Steel*. When sedentary, domestic farming, what Diamond calls "farmer power," becomes the sole food production system in a given society, as it inevitably does, it generates population growth and then an urbanized technology that eventually enslaves or annihilates hunting and gathering cultures with which it comes into contact. Ironically, farmer power then enslaves the farmers who generated it. Farmer power subsequently leads, as Diamond so analytically argues, to rapid progress in manufacturing, to an elite science, to an organized bureaucracy, to military might, to an oligarchy of wealth, to an influx of germs and disease and, ultimately, to decline. For security, the power structure amasses large supplies of annual grain. Because its concentration of nutrients are in a form that can be stored, shipped, traded, and sold far from its place of origin, annual grain engenders a most enticing, if eventually false, fortress against the fear of starvation and subsequent loss of power. It becomes the coin of the realm, breeding almost insuperable political consolidation. It did so for the ancient Egyptians and the Romans and the Mayans for awhile, and now, for awhile, America gets its turn.

Could humankind for once end an old cycle and begin a new one based on a pastoral food system without an intervening decline or collapse of the economy and civilization? As a reader of history, I can't find it in me to be optimistic. Just hopeful. If we can master the art and science of grassland farming as we have mastered industrial grain production, and then allow real economics to lead the change, unfettered by the political or financial interests of the status quo, we might pull off a first in history. But I can't see "farmer power" resisting the temptation to manipulate economics to its own ends, as it is now already doing, and unwittingly accelerating the collapse. If that problem were avoided however, we might gain, without upheaval, something more beneficial to humanity than farmer power's boast that a few big businesses can "feed the world." I dare only whisper to the bees humming in the clover what that something could be: a paradise of permanent pastures.

Notes

1. Masanobu Fukuoka, *One Straw Revolution* (Emmaus, Penn.: Rodale Press, 1978), 15.

2. H.D. Hughes, Maurice E. Heath, and Darrel S. Metcalfe, eds. *Forages* (Ames: Iowa State University Press, 1951).

10

The Uses of Prophecy

David W. Orr

For nearly four decades Wendell Berry has written about farming, soil, nature, and community without ever becoming repetitious or boring. He is an agrarian, or more accurately, the preeminent agrarian. From Hesiod to the present no one has represented the agrarian cause with greater eloquence, logic, or consistency. The power behind the writing, however, is the close calibration between his words and the life he's lived—"a principled literary life" as Wallace Stegner once put it. If there are discerning readers two centuries hence, I have no doubt that Wendell Berry's novels, essays, and poetry will still be read. But there is cause to doubt whether the agrarian philosophy of Wendell Berry will be more than a footnote in a history of technological expansion, urbanization, and economic growth until things come permanently undone. He is widely admired for his literary gifts and wisdom, yet much dismissed. Why is this so? What does it say about who we are and about our larger prospects? And what might be done about it?

Interpretations

The most common reaction has been to concede his literary talents while dismissing the substance of what he has to say as nostalgia for a bygone, but better, time. Many can recall a time of prosperous farms and rural communities, places that fostered good children, decent citizens, competent people, land stewards, and real patriots. The tens of thousands of visitors to Amish country in Ohio and Pennsylvania are daily testimony to the appeal of a life rooted in the land and in stable rural communities. For many this is not nostalgia, but an awareness that in some places and at some times people did get the relation between culture and land right, and that remembrance haunts the modern mind. Jacquetta Hawkes, for example, once described rural England of the eighteenth century as characterized by a "creative, patient, and increasingly skilful love-making that persuaded the land to flourish."[1] Such times and places were not perfect by any means, but they did represent an exceptional quality of life. That such places are now mostly confined to memory is a fact to be lamented but not undone because of changes in farming, culture, and society. More importantly, the small farms favored by Wendell Berry would be inadequate to feed a growing world population. The necessity imposed by sheer numbers requires the industrialization of farming, which means large amounts of capital, large-scale farms, chemicals, food processing, and companies like ADM and ConAgra to feed the world. Others like James Lovelock arrive at roughly the same conclusion by a different route. Agriculture, they say, must be replaced by a technology-intensive food system in order to free land for wilderness, store carbon out of the atmosphere, preserve biological diversity, and stabilize the earth's vital signs.

Second, most would say that the benign world described by the agrarians never truly existed. Farming and rural life were boring, hard, insecure, and sexist. The agrarian world, furthermore, was rooted in injustice because the land, here and elsewhere, was stolen from natives. Farm life was difficult and economically insecure. The agrarian world was dominated by men to the great disadvantage of women, children, and nature. The demands of farm work drove out the development of other talents and interests so that human potentials and real culture withered on the vine. Rural communities were often closed and hostile to social, religious, and cultural diversity. Agrarian communities, closed to outsiders, were "provincial" in outlook. At their worst, they were the violent, evil, and confining places described by William Faulkner. Agrarianism? Good riddance!

Agrarians must acknowledge that this indictment has some truth, but it is not the entire story. Farming has always meant hard work and economic insecurity compounded by exploitation at the hands of the crown, the nobility, or the church. In our own history, agrarianism flourished in New England for a time and in parts of the Midwest but is now most evident in Amish communities. Its decline can be attributed to many things, including the fact that many farmers, as Berry says, did not come to stay but to make money and cashed out at the first opportunity. For the majority, however, the relentless pressure of the market undermined care for the land, cooperation, thrift, community, and permanence. In other words, most of what we know about farming and farm life is how it was done and lived out in situations where farmers and land alike were greatly exploited.

A third reaction to the agrarianism of Wendell Berry is to say that it is just not in our nature to scratch the soil, sow, reap, and live in settled communities. Agriculture is a ten-thousand-year-long aberration caused by overpopulation and the necessity to feed urban masses; a wrong turn in human evolution. Paul Shepard, for one, argued that humans are really hunter-gatherers and that tending the soil was a late and unfortunate overlay on our true nature. A better society, he argued, would be oriented to a kind of high-tech hunting and gathering. But how are we to reenact the hunting-gathering ideal in a world organized along very different lines for the past ten thousand years? The only plausible answer is that we cannot.

A fourth view holds that agrarianism whether good, bad, or otherwise was simply a stage in human evolution that has passed because our technological prowess is taking us in other, and presumably better, directions. Food might soon be grown without anything like present-day farms and farmers, or even soil. It is possible to create foodlike substances from oil in factories that require much technology and few people. Taking this a step further, professor Hans Moravic proposes that we even shed the body and "download" our minds into computers attached to machines that function like bodies with arms and legs.[2] Since machines can be repaired, we would attain immortality of sorts. But assuming this is possible, what exactly would "we" be? Surely we would not be human in a way recognizable to us now. Our human nature is the result of a long conversation between the body, mind, spirit, culture, and physical nature.

Mind is not separable from the body as Descartes and Moravic would have it. Nor is the "embodied mind" separable from the nature of places. The improvers, such as Moravic, intend an end run around mortality and with it our dependence on soil, rain, and the skill of the farmer. Food is replaced by the need for electricity. At mealtime one would plug in, not dine. More conveniently, perhaps our "progeny" could be engineered to photosynthesize directly with minds engineered to believe this to be an improvement. Whether Moravic's world could be made to "work" I do not know, but I do know that it represents a point of no return.

A fifth reaction to Wendell Berry is to say that in the best of all possible worlds we would live in predominantly agrarian communities as Berry proposes, but we aren't up to it. As Dostoyevsky's Grand Inquisitor in *The Brothers Karamazov* puts it to a silent Christ, humanity is "weak, vicious, worthless, and rebellious . . . in the end they will lay their freedom at our feet, and say to us, 'make us your slaves, but feed us.'" Christ, in Dostoyevsky's telling, remains ambiguously silent. However said, this is the view of all of those who purport to be a "supermarket to the world" or to "do it all for you" without any effort, skill, or even awareness on our part. There is no crime but hunger, as Dostoyevsky says, so whatever it takes to feed the world is permissible. Given the chance, we would trade the toil and insecurity of agriculture for the promise of ease, abundance, and security. And once the trade is consummated, we become accustomed to conveniences, luxury, and, as Wallace Stegner once put it, all of those "things that once possessed could not be done without." Farming, in other words, is a burden that most farmers discard as soon as they are able. That view is rendered plausible by the history of farm technology from the middle of the nineteenth century to the present. With the notable exception of the Amish, farmers adopted virtually every device that promised higher production with less physical effort. In the process, farming as a way of life gave way to agribusiness. In other words, there are good circumstances that we cannot sustain and visions that we cannot live up to.

Victor Davis Hanson believes this to be the fate of small-scale agriculture by which the small farmer is doomed: "The sad history of complex societies, ancient and modern, argues that bureaucracies grow, never shrink, and so suggest that these futurists—not agrarian romantics—have seen the real forecast over the horizon. Unproductive citizens multiply, rarely wane. Taxation, urbanization, and specialization are the harvests of elite legitimizing and nuancing classes—government, insurance, advertising, law, finance—who feed and clone from ritual, regulation, and regimentation."[3] The agrarian world is being crushed by the remorseless development of civilization and by the lack of courage to return to "a common culture of peers, life away from the vortex of pelf and publicity, a firmness with the poor as well as the wealthy, an embrace of rural shame, a rejection of convenient urban guilt" (274–75). Hanson offers no hope for the future family farm: "We are now in the third stage of a future that has no future, an agrarian Armageddon at the millennium where the family farm itself . . . will be obliterated" (xi).

In some places, however, stable agrarian communities existed for many centuries. When they eventually declined or disappeared, it was not because rural people were weak, lazy, or incompetent. To the contrary, they mostly failed because of larger economic and political forces. Historically, European peasants were bled white by taxation. The English peasantry were forced to the margin by the enclosure of common lands and by the political power first of the landed nobility and then by that of industrialists. The prosperous Russian peasants (kulaks) were coerced into collective farms by Stalin. In our own history, the decline of farms and farm communities is a complicated story of taxes, land prices, government policies, and the treadmill of technological improvement. As a result, throughout the twentieth century many left farming more from necessity than choice. The deck was stacked against the farmer, and particularly the small farmer. And that is not all.

Whatever status farmers once had has withered in the onslaught of the advertising, entertainment, and communication industries. Agrarian virtues of honesty, thrift, practical competence, and neighborliness had no place in the glittering, fast paced, consumer-oriented world of Madison Avenue and Hollywood. A way of life dependent on soil, hard work, ecological competence, and devotion to place became a source of shame. Children of farmers could not help but compare their parents with the slick images of the smart city people living effortless, exciting lives as portrayed in magazines, catalogues, and the movies. But this is not the kind of individual weakness described by Dostoyevsky as much as it is evidence that we are vulnerable to the considerable powers of modern psychology and communications technology deployed to make us dependable customers. Advertising and entertainment industries have become adept at selling a life "style" requiring lots of cash and an agreeable willingness to part with it on a whim—further evidence of the corruption of society organized around the logic of finance capital and exploitation.

Finally, Page Smith had it right, I believe, when he described Wendell Berry as "*the* prophetic American voice of our day." This perspective may help to explain why he is both widely admired and often ignored. We don't much like prophets because they make us feel uneasy. They see things most prefer not to see, and say things many wish went unsaid. They are ambiguous figures who point out the gap between an unhappy reality and better possibilities that reveals us for what we are. Instead of being forgiving and therapeutic, prophets are, as Abraham Heschel once wrote, "impatient of excuse, contemptuous of pretense and self pity."[4] But prophets do not just condemn, they intend to move us toward better possibilities. They call to mind a time when we were better people, but they also look forward to a time when we might be restored to some semblance of grace, which is to say that they are accusatory and forgiving. Prophets are poised between the past and a better future. To dismiss Berry as simply nostalgic, in other words, misses the point. He ought to be read as much as a futurist describing better possibilities as someone looking back to what once had been. To argue that his view of the past is mistaken is, I think, to misread both Berry and the historical record. Decent agrarian communities have existed and their various imperfections do not diminish that fact. To dismiss agrarian possibilities that Berry describes on the grounds of necessity is to surrender our ability to choose to an unwarranted and unworthy technological, demographic, or economic determinism. To propose that we further technologize the food system or shed the body for some machine-like existence is sheer madness. But whether we are up to the challenge of going forward to an authentic agrarian future is another matter entirely.

The Powers of Denial

The obstacles to agrarianism of any sort are many, beginning with the obvious fact that we have built a massive infrastructure of pipelines, power plants, roads, bridges, airports, railroads, mines, factories, shopping malls, and so forth around the idea of permanent economic expansion powered by cheap fossil fuels. Much of that infrastructure is dangerous, vulnerable to terrorists, polluting, and obsolete. A sizeable portion of it is in need of repair and must soon be replaced. More important, industrialization is also embedded in our minds and limits our ability to imagine better possibilities. Industrialization rests on the simple and seductively powerful idea that we can exploit soils, forests, biological diversity, and minerals without adverse consequences, and that doing so is akin to our rightful destiny. That idea is widely known to be wrong, even perversely so, but it still exerts a powerful hold on the public mind and public policies. Some find it difficult to consider the possibility that paying the full social and environmental costs would have radically changed the scale and scope of industrialization.

Agrarians, however, have always understood this. They give priority to honest accounting that includes soil, land, community, good work, and to *agri*-culture as the foundation for culture as Berry puts it. They believe, in Liberty Hyde Bailey's words, "that the farm and the backspaces have been the mother of the race" because "they beget men and women to be serious and steady and to know the value of every hour and of every coin that they earn; and whenever they are properly trained, these folk recognize the holiness of the earth."[5] If that sounds quaint to some, the advantages of a new kind of agrarianism in the twenty-first century—considering the alternatives—are increasingly compelling to others. Concern to preserve farms and farmland is evident in recent citizen initiatives across the country. The market for organic food is rising by double digits each year. Gardening programs exist in virtually every major city and in many schools. The sustainable agriculture movement, if not yet a significant political force, has grown steadily for three decades. And after the events of September 11, 2001, there is a powerful case to be made for a new agrarianism to secure the nation's food supply. The resilience that once characterized a distributed network of millions of small farms serving local and regional markets made it invulnerable to almost any conceivable external threat, to say nothing of the other human and social benefits that came from communities organized around prosperous farms.

That the political, economic, intellectual, and ecclesiastical leadership of the country largely ignores such things says nothing about the wisdom of a new agrarianism and everything about the concentration of political, economic, and intellectual power, and the hold that industrialism still has on the public mind. What passes for farm policy, for instance, is little more than a vast system of subsidies that enriches agribusiness and corporations, while promoting overproduction and driving small farmers out of business here and elsewhere—all justified in the name of economic rationality. For agrarians the deck is stacked higher than ever. Why?

The answer is not simply that we have built a massive industrial infrastructure and therefore have no other options. Rather, it is that we continually fail to implement better possibilities. We must now ask why. What stops us from making an effective and rapid transition from fossil fuels to efficiency and renewable energy? What stops us from pricing resources at their true costs? What prevents us from conserving farms and open spaces? By what logic do we fail to account for the costs of climate change or pollution? Why do we subsidize overcutting in the national forests or give minerals on public lands to large corporations for a pittance? Why do the already wealthy grow ever richer and the poor ever poorer? None of this is inevitable, but is a kind of willing sleepwalk toward the edge of an abyss. The industrial world, in other words, is not just a physical reality, it is a system of denial.

A large part of that system has to do with the corruption of our democracy by the largely unchecked power of corporations and the hugely wealthy. America is fast becoming a colony controlled by a small class of exploiters who have no allegiance to the land or democracy, and no sense of responsibility beyond that to maximize short-term shareholder value. They consider themselves to be a part of a global economy and no part of any particular place or local economy. Power, accordingly, has gravitated from citizens to corporations who do not consider themselves citizens at all and whose lobbyists diligently troll the halls of Congress and state legislatures to purchase votes, access, and absolution. Recent revelations about corporate accounting scandals are just that: recent revelations—the small tip of a very big iceberg. The pattern of abuse, which has been there all along, is once again exposed for what it is and has always been: a system seldom beholden to good public purpose and consistently aimed to further enrich the wealthy. And the public, once again, are given solemn promises of reform by public officials who are themselves indentured to the money class. A mendacious media, a fully owned subsidiary, duly reports such things between its advertisements. None of this is news to readers of Wendell Berry or to the merely alert.

Neither would it be new to say that denial and political corruption requires the corruption of language. By some strange alchemy, the word "conservative" has been co-opted by those intending to conserve nothing except the rules of the game by which they are greatly enriched. That they wish to bamboozle should astonish no one; that they get away with it, however, depends on a high level of public drowsiness and gullibility. But that is an altogether more complicated thing—a kind of co-conspiracy involving a combination of ignorance and apathy on one side and a desire to mislead on the other, all disguised by a language unhinged from reality. We come to believe that "Coca-Cola is the real thing" and, by a similar logic, that G.W. is a real president as were, say, Lincoln or Roosevelt. Perhaps we assume that none of this matters very much anyway because images are what count, not place, tradition, obligations, physical reality, or our long-term prospects. We are complicit in the corruption of language and in the very process whereby we are being dumbed down. Language, as George Orwell once put it, "becomes ugly and inaccurate because our thoughts are foolish, but the slovenliness of our language makes it easier for us to have foolish thoughts."[6] And foolish thoughts we have in surplus, goaded on by television and the $300 billion spent by the advertising industry to promote foolishness.

One might expect that sloppy language would stand revealed as nonsense by the rigorous power of numbers. Alas, numbers in the service of denial are corrupt as well. One need look no further than the fictions offered up by economists that exclude biology, ecology, morality, and thermodynamics. The practice of capitalist bookkeeping has no debit column in which we would subtract the loss of soil, forests, water quality, wildlife, and countryside. What counts, as they say, is what can be easily counted. The "services of nature"—including wetland water filtration, soil and temperature stabilization, and water retention—more elusive but no less real, are neglected, but that does not thereby make us wealthier. At best that neglect deceives and at worst it conceals the transfer of wealth from country to city and from future generations to the present, which is to say it is theft.

Denial does not end with politics, advertising, or economics; it has invaded the world of research and education. In *The Unsettling of America*, Berry describes the perversion of the intentions behind the creation of land grant universities by which they have become adjuncts to agribusiness. The problem has grown worse in the decades since its publication. Science, as Berry argues in *Life Is a Miracle*, has become increasingly absolute in its reach and in its pretensions of omniscience. New developments in genetic engineering, nanotechnology, and artificial intelligence threaten to take us beyond a point of no return, yet we seem unable to act with foresight. Intelligence, often confused with the encumbrance of learned degrees, is conspicuous by its absence in debates about science and the proper boundaries of technology, and this, too, has to do with an older corruption of language and words such as "progress" and "development." Educational and research institutions were fashioned as industries that in time became useful and beholden to other industries in the faith that they thereby served human progress and development without much thought about what those words meant. In time, I think they came to mean something far removed from what was once intended—an escape from the human condition itself.

This leads me to a final observation. Agrarian life places people in close contact with the cycles of birth and death. On the farm, something is always being born and growing or being led to slaughter and dying. Death is simply part of a natural cycle that farm children experience early on. That proximity to death as an everyday occurrence doesn't make it easier, necessarily, but it may make it easier to comprehend and accept as part of the natural order of things. The closeness to birth, growth, decay, and death in the nurturing context of a farm has the effect of demystifying mortality and laying the psychological foundation for the healthy acceptance of our own death. The demise of small diversified farms and people in contact with farming, then, is not just a change in the way we produce food and organize the countryside, but a change in how we think about the basic facts of living and dying. Industrial agriculture conceals death on a massive scale, making it "efficient" and thereby ugly and sacrilegious.

Presently, we are witness to death on the largest scale imaginable—that of life on the earth itself. This, too, is concealed by the logic of efficiency and is a sacrilege on a scale that we have no power to describe. No day passes without news of the decline of species, seas, forests, lakes, rivers, all spiraling downward into the destabilization of the planet's biogeochemical systems. We seem paralyzed by that fact, or perhaps strangely fascinated by it. Paradoxically, we are causing death at the largest scale possible, yet no culture has ever taken greater pains to deny mortality or spent more of its treasure to ward off the mere appearance of impending mortality. It is plausible that no previous culture has had greater difficulty coming to grips with death, our fears magnified by our technology. We flirt with death as voyeurs—fascinated by the violence and death portrayed on film and television. Our children grow up playing violent video games. The abuse of drugs and alcohol is widespread. A few participate in extreme sports but most of us drive carelessly and knowingly take many other risks. Our minds are populated with figures such as Hannibal Lector and Jason of *Friday the 13th* fame, and any number of serial killers whose faces are featured on trading cards like sports figures. Death terrifies and fascinates us as it has no other people. Teenage boys, in particular, have a fascination with death, but few culturally approved and healthy ways in modern society to work through that stage of life to a stable maturity. How are we to overcome our need to deny death, and what would this mean for the larger society?

The psychologist Ernest Becker spent much of his life trying to understand why we deny death, what psychologists now call "terror management," and how we do so. Denial of death takes many forms: "heroic transcendence, victory over evil for mankind as a whole, for unborn generations, consecration of one's existence to higher meanings."[7] Toward the end of his own life Becker concluded that "we are living the grotesque spectacle of the poisoning of the earth by the nineteenth-century hero system of unrestrained material production. This is perhaps the greatest and most pervasive evil to have emerged in all of history, and it may even eventually defeat all of mankind."[8] But he believed the causes of the evil to be ironic: "Men cause evil by wanting heroically to triumph over it, because man is a frightened animal who tries to triumph, an animal who will not admit his own insignificance" (151). The tragedy "is that [evolution] created a limited animal with unlimited horizons" (153). Increasingly we manage our terror of mortality by trying to extend our power over nature through heroic feats of science, technology, and economic growth.

Life in an agrarian community, in contrast, is not heroic, nor is the denial of death a particularly useful response for farmers. Agrarian life, rather, requires a patient and painstaking accommodation to the realities of life and death in the effort to husband the health and long-term productivity of particular places. And this fact may help to explain some part of the rejection of family farming and agrarianism in our time. Good farming, as Wendell Berry describes it, does not lend itself to heroic projects, technological fantasies, denial by consumption, carelessness, or great celebrity, requiring instead qualities of steadiness, hard work, neighborliness, practical competence, thrift, and perseverance. It is not far-fetched to believe that we are now in full flight from such qualities, hoping to escape our bonds to mortality, time, work, nature, and our own nature. Neither is it far-fetched to believe that this flight has been aided and abetted by the same people and the same institutions that have corrupted our politics, economics, language, and education.

POSSIBILITIES

Berry describes two general organizing principles for society: one industrial, the other, agrarian. The logic of industrial capitalism has brought us to a cul-de-sac. Instead of the limitless abundance promised, we face ruin on a global scale and as far into the future as the mind can comprehend. There are those who believe, nonetheless, that we might make a second and better industrial revolution. As proposed, that revolution requires no improvement in our politics, desires, economic thinking, connections to the land, and moral philosophy, only greater cleverness in how we handle materials so that we generate no waste and live on sunlight. There is much that is useful in this, but in the end it will founder on the shoals of hard reality and human recalcitrance. Why would people much enamored of wealth, convenience, and consumption decide to rouse themselves to heroic feats of cleverness especially if the reward is not theirs? By what logic and by what means will capital rigged and outfitted for short-term exploitation of land and people be reformed to husband people as well as the "natural capital" of soils, forests, waters, and biological diversity over the long term? In what way will people having defined themselves as consumers, and thereby having diminished their role as a public, decide to work for a common good including that of nonhuman nature? I cannot imagine a system built on exploitation, consumption, growth, and uniformity—however cleverly managed—as anything other than a prelude to ruin. To say that industrialism, even a smarter version of it, cannot be sustained is not, however, to say that agrarianism will succeed. Darker possibilities are certainly imaginable. Perhaps agrarianism will prevail, as Churchill once said of democracy, only after we have tried everything else—if we have not ruined everything that really matters in the trying.

What, then, is the future of agrarianism? Is it possible, as Eric Freyfogle hopefully asserts, that even in a rapidly urbanizing world "agrarianism is again on the rise"?[9] If so, is it also possible for that rise to become a flood that recasts the industrial world in an agrarian mode? How might this happen? Few good social changes occur without hard work, organizing, and effective strategy. Agrarians are no strangers to the first, but are famous neither for organizing or thinking strategically. They are better known for their independence, self-reliance, and, yes, crankiness. The Southern Agrarians, for instance, having delivered themselves of their manifesto, *I'll Take My Stand*, mostly disappeared back into their various writing and academic endeavors, and the world went on as before. The sustainable agriculture movement of the past three decades, for all of its accomplishments, still exists at the margin of our politics. The power of agribusiness, petro-chemical companies, and the food industry seems undiminished.

Still, I believe Freyfogle is onto something for reasons that Wendell Berry has explained in detail for nearly four decades: we have no better alternative. But that is not to say that agrarianism is inevitable everywhere or in any particular place indefinitely. To the contrary, the making of an agrarian world will require a great deal of thought, effort, political savvy, stamina, and orneriness. This is not about tinkering with "the system," but a more fundamental change in how we relate to the land and to each other not as exploiters but as members of a community who intend to stay.

Agrarianism, as described by Wendell Berry, is no small, whittled-down philosophy for rural folks. It is, rather, a full-blown philosophy rooted in the realities of soil and nature as "the standard" by which we also come to judge much more. It is grounded in farming, but is larger still. The logic of agrarianism, in Berry's work, unfolds like a fractal through the divisions and incoherence of the modern world. It is, in his word, a "re-membering" of the wholeness and the Holiness of the creation. His is a philosophy that begins with place, soil, and farming, but is extended to include race, religion, sexuality, science, politics, wilderness, economics, world trade, food, foreign policy, and more. For the first time in the long history of agrarianism we have a philosophy that doesn't end at the farm gate with a description of the bucolic pleasures of tending to the soil. Berry's great achievement, I believe, is to describe eloquently and in great detail how our connections to soil, food, and agriculture extend through virtually everything else. He's given us a grounded philosophy of the wholeness of things with the admonition to "solve for pattern."

The pattern includes much more than farming, although that is the starting point. For agrarianism to work, it must have urban allies, urban farms, and urban restaurants patronized by people who love good food responsibly and artfully grown. It must have people who appreciate the pleasures of eating and who regard it, as Berry has said, as a political act. It must have farmers who regard themselves as trustees of the land that is to be passed on in health to future generations. It must have communities that value farms, farmland, and open spaces. It must have visionary and courageous policymakers who serve as trustees for our common wealth. It must have scientists, like Wes Jackson and his colleagues at the Land Institute in Kansas, who will create the knowledge necessary for a twenty-first-century agrarianism. It must have bookkeepers that tell us the truth about wealth and its limits. It must have elders to remind us that we, the soil, and all that lives are part of the same ancient pattern, one and indivisible. This is only to say that an agrarian world requires a discerning public that understands that health, too, is indivisible.

This public cannot be willed into existence, it must be educated to regard itself as a public and to understand the connections between food and the health of the land, soils, and waters. But here's the rub: the practices, traditions, and memories of farming have been passed down from one generation to the next in the daily routines of living on the land. They have seldom been an important or effective part of any formal curriculum. To the contrary, schooling has mostly been thought of and defended as a way to escape the farm and to improve one's economic possibilities by becoming useful to the industrial economy. For agrarianism to succeed, I believe that this must change and, in fact, is changing. But much more will have to be done.

A public that understands the relationships between soil, forests, water, food, and health requires an educational system that equips students to comprehend systems and patterns. Every high school graduate ought to understand the connections between the dead zone in the Gulf of Mexico, farm practices in the corn-soybean belt, the depletion of the Ogallala aquifer, oil wars, the rising tide of obesity, dying rural towns, urban sprawl, and antibiotic resistance. But this in turn would require a curriculum no longer organized exclusively by disciplines, nor one that rests on the assumption of human dominance over nature as a matter beyond debate. It is a curriculum that enables students to think "at right angles" to narrow disciplines, as Aldo Leopold once put it, and one that fosters ecological imagination. But that is a transformation well beyond what typically passes for educational reform.

At the earliest ages, education ought to do the work once done at home of connecting children to their places. School gardens and gardening could be a cornerstone of the daily experience of children. The Center for Ecoliteracy in Berkeley, for example, is helping to establish school gardens in public schools throughout California as a way to improve cafeteria meals, teach biology, earn money from the sale of the surplus, and add to children's sense of connection and competence. The values of gardening are many; as an educational tool they can help to develop the ecological imagination of children in association with pleasurable activity. The same holds true at higher levels as well. For a dozen years I have taught a course in sustainable agriculture to students at Oberlin College with mostly urban backgrounds. Many of those students have participated in the creation of a community supported farm, restaurant, and health food store. The strong interest here and on other campuses reflects a great desire to connect with the land and reform the food system along lines that fit agrarianism through and through.

Further, the buying power of schools, colleges, and universities ought to be used to support local farms and farmers. A dependable market for locally grown farm products can be an important stimulus to help establish sustainable agriculture. Students at California State University at Chico, for example, purchase T-shirts made from organically grown cotton to sell in the campus store. And there are more ambitious possibilities as well. Colleges might use a small portion of their assets to purchase farms that would otherwise be developed, leasing these back to young people wanting to farm but lacking the financial means to do so, thereby maintaining open space and protecting the local heritage, environment, and long-term values. Further, colleges and universities could provide a guaranteed market for the produce from such farms and help establish relationships with older farmers who could serve as mentors in the practical arts of agriculture. With some imagination, in other words, they could help to stop urban sprawl and reestablish viable local farms.

And there is one other large possibility. The cornerstone of the industrial world rests on the availability of cheap fossil fuels and the belief that it is our right to burn them as we please. An agrarian world must be powered on contemporary sunshine. But in the present vacuum of leadership on climate policy, schools, colleges, and universities could lead in the transition to the solar age. To that end many colleges are beginning to develop plans to become climatically neutral in the next few decades by improving efficiency, buying "green power" from utilities, and applying advanced technology such as photovoltaics, fuel cells, and wind turbines. The creation of a distributed energy system, similar to distributed computing, would reduce vulnerability to terrorism, eliminate emission of greenhouse gases, lower our deficit, improve energy efficiency, and build a citizen-based energy system. It is no longer a question of technology, but one of leadership and vision.

There is serious work to do to elaborate and extend the agrarian idea and overcome the biases inherent in a curriculum organized around "upward mobility" and economic growth. Books such as this one will help, but the possibilities are many. Agrarianism could fail in the public marketplace because of the lack of coherent, well-articulated, and forcefully presented ideas, however true and important. Another Berry, Thomas Berry, calls all of this "The Great Work." Wendell Berry proposes that we aim to do great work in the minute particulars of living within our means as persons and as a people in the full awareness that the world is rich in possibilities. What use will we make of those prophetic possibilities?

Notes

1. Jacquetta Hawkes, *A Land* (Boston: Houghton Mifflin, 1950), 202.
2. Hans Moravic, *Mind Children* (Cambridge: Harvard University Press, 1988).
3. Victor Davis Hanson, *Fields without Dreams* (New York: The Free Press, 1996), 282.
4. Abraham Joshua Heschel, *The Prophets: An Introduction* (New York: Harper Torchbooks, 1969), 7.
5. Liberty Hyde Bailey, *The Holy Earth* (Ithaca: New York State College of Agriculture, 1980), 27.
6. George Orwell, *A Collection of Essays* (New York: Harcourt, Brace, Jovanovich, 1981), 157.
7. Ernest Becker, *The Denial of Death* (New York: Free Press, 1973), 268.
8. Ernest Becker, *Escape from Evil* (New York: Free Press, 1975), 156.
9. Eric T. Freyfogle, ed., *The New Agrarianism* (Washington: Island Press, 2001), xiii.

Part 3

Putting Agrarianism to Work

11

Country and City

The Common Vision of Agrarians and New Urbanists

Benjamin E. Northrup and Benjamin J. Bruxvoort Lipscomb

> The question before us, then, is an extremely difficult one: How do we begin to remake, or to make, a local culture that will preserve our part of the world while we use it?
>
> —Wendell Berry

Since the end of World War II, Americans have been engaged in a great experiment: the reconfiguration of their built environments around the automobile. Until then, humans had built cities and towns at the scale of the pedestrian. A short walk was the constant measuring rod in Mesoamerican capitals, Roman encampments, old European cities, and New England villages. Following World War II, however, Americans began to reshape, and often replace, their established communities to make room for automobiles. In order to accommodate these devices, they had to enlarge everything, to build a world for giants. This vast project included the invention of the interstate, the shopping mall, and tract housing. Meanwhile, middle-class Americans fled established cities and towns in pursuit of freedom, open space, and a new start. Americans have not, for the most part, thought of all this as an experiment. They have regarded it simply as progress. But as the once-vibrant Main Streets of America have lost out to sprawling commercial strips, and pleasant Sunday drives have been replaced by long, obligatory commutes, some are asking whether there is another way.

To agrarians, this story will seem strikingly familiar. Agrarians note that modestly scaled, largely self-sufficient farms—a standard type throughout history—declined precipitously after World War II, a decline that closely paralleled that of our cities and towns. Tens of thousands of such farms have now failed; they have failed, indeed, at faster rates than they did during the Great Depression. These small, diversified farms have been replaced by large monocultures, the agricultural equivalent of tract housing. The loss of these farms has brought on a depopulation and, consequently, a degradation of the landscape that parallels the depopulation and degradation of our old cities and towns. Just as urbanites were seduced by the freedom and mobility promised by the automobile, farmers were seduced by the ease and efficiency promised by new machinery and chemicals.

It is clear that country and city share a history of recent decline. Are they likewise dependent on each other for their recovery? Agrarians have long observed that the eating habits of urbanites have an enormous impact on the viability of good farming, and they have counseled urbanites to conform their lives to a pattern consistent with good farming. Wendell Berry says this with characteristic concision: "Eat responsibly."

But surely there is more to say. Agrarianism is a comprehensive political and cultural standpoint, aspiring to address both rural and urban contexts. Our main task in this essay is to chart a largely unexplored confluence of views between agrarians and the increasingly large and prominent group of architects, elected officials, and concerned citizens advocating the New Urbanism, a traditional approach to urban design. As we will show, there are many similarities between the views and goals of the two groups, and some important lessons to be learned from the encounter.

New Urbanism

Both "new" and "urbanism" may have a suspicious ring to agrarians, for whom "new" has usually meant the consolidation of land ownership and the dispossession of small farmers, while "urbanism" connotes the sprawling expansion of American cities. New Urbanists, however, share the agrarians' disdain for contemporary patterns of land use, taking inspiration instead from traditional models. New Urbanist practice is frequently referred to as traditional neighborhood design. Indeed, the urban forms that New Urbanists advocate are generally not new. What is "new" is the context in which New Urbanists work to apply their ideas.

America's experiment with the automobile has virtually put an end to traditional ways of building cities and towns. Jane Jacobs was among the first to question the impact of our new automobile-dependent approaches in her book *The Death and Life of Great American Cities* (1961). Her work, and that of Luxembourg architect Léon Krier, inspired a generation of architects to challenge the new pattern of human settlement. The reaction was not merely negative, however. The new generation found inspiration in established neighborhoods and towns across the country, in the narrow streets lined by trees and front porches, in the courthouse squares of small towns, in the vast, picturesque parks of great cities, in the ubiquitous corner store, and in the apartments located above it. The very elements that had been discarded after World War II now seemed charming and functional. Ironically, it was the onslaught of automobile culture that prompted architects to reflect on and articulate the structural patterns that had undergirded human settlement for centuries. These architects, the instigators of the New Urbanism, concluded that human settlements have traditionally been oriented toward the pedestrian and, furthermore, that the neighborhood is the fundamental unit of human settlement.

This latter proposition may seem strange, for is not the city or town—the complete local community, however defined—more fundamental to one's identity than one's neighborhood? In a political sense, it likely is. But in terms of formal and practical structure, New Urbanists have observed, all traditional human settlements, from cities to villages, are made up of similar atoms: neighborhoods. Based on observation of existing conditions in established towns and cities, New Urbanists define the neighborhood as a walkable district with a mix of housing, commercial space, and civic buildings. It is characteristically a five-minute walk from the center of a neighborhood to its edge.

A city, traditionally conceived, is a large congregation of neighborhoods. But as we have just observed, neighborhoods exist in other contexts. Though we do not normally think of a village this way, it is essentially a single, freestanding neighborhood in the countryside. A town is composed of several neighborhoods, whether they are called that or not. A city is a large town, comprising many neighborhoods, each self-sufficient to a degree, but each also reinforcing the others.

It goes without saying that neighborhoods vary greatly: in context, in economic structure, in demographics, and in other ways. But as a group, they also have much in common. Good traditional neighborhoods typically have discernible centers and edges. Within the geographic bounds of a neighborhood, residents have traditionally been able to meet most of their daily needs. The offices and shops where such goods and services are provided are located at or near the neighborhood center and are, as already remarked, within a short walk of all neighborhood residences. The housing stock is sufficiently varied to accommodate people at different stages of life. Neighborhoods also feature parks, civic buildings, schools, and other public places that foster and symbolize communal identity. While some neighborhoods are denser than others, all traditional neighborhoods have sufficient density—and the necessary structure of walkable, interconnected streets and blocks—to support a vibrant street life. In no case is overall density as low as in contemporary suburbs, which cannot support an active street life no matter how many sidewalks and parks are present.

Perhaps the most important feature of traditional neighborhood design is its mixture of uses: commercial, residential, and civic. Obviously, a neighborhood in the center of a city will have a high proportion of commercial and retail space. A neighborhood in a small town or an old streetcar suburb will have a preponderance of single-family detached houses. But both of these neighborhood types contrast sharply with what we find in contemporary cities and postwar suburbs. These settlements are governed by zoning laws that enforce segregated uses, so that residential neighborhoods are forbidden to have corner stores, and commercial zones can have no apartments.

Zoning was invented in the 1920s in response to a real problem: noxious factories locating near residences. But after World War II, concurrent with the emergence of an automobile-centered culture, single-use zoning was zealously applied everywhere and to everything. Residential, commercial, and civic uses were distinguished from one another, and increasingly fine distinctions were made within each category. Generally speaking, the result of each new distinction was that uses became more and more segregated. This effectively made the traditional neighborhood, in all its diversity, illegal. It also severely compromised the quality of life of all citizens too young, too infirm, or too poor to drive, nearly half the population.

The details of neighborhood design matter, too. And New Urbanists have a host of recommendations about the block, the street, and the building. These small-scale recommendations are, again, grounded in their observation of traditional neighborhoods and their desire to revivify the public realm. Streets, they urge, should be lined with sidewalks to encourage pedestrian traffic. In most areas, the sidewalk should be separated from the street by shade trees or parallel parking, which act as buffers between the pedestrian and automobile traffic. Blocks should be small enough to allow pedestrians to navigate the neighborhood easily and to dilute automobile traffic by spreading it onto multiple routes. Residential and commercial buildings should face the street and have clear entrances; they should also be close enough to the street to create a sense of spatial definition, giving it the feeling of an outdoor room. And buildings should feature windows facing the street to make it feel inhabited. Civic buildings, such as libraries, schools, and churches, should receive prominent sites; they are allowed to break the architectural rules governing other structures so they will stand out as buildings of special importance.

Similarly, New Urbanists have a variety of suggestions for design at the regional scale. Many of these resonate with agrarian concerns. Those who love the traditional city regularly call our attention to its distinct boundaries. Most European cities still have an edge where the city stops and fields begin. Truck farms on the edge of these cities have traditionally provided fresh, locally grown produce. American cities used to conform to a similar pattern, but over the past fifty years, their edges have been obscured by sprawling development. The contemporary suburb is neither city nor countryside, but an uncomfortable fusion of the two. The rural landscape that suburbia was intended to procure for the city dweller has been lost. As farms and other undeveloped spaces disappear, we are left with a strange landscape. It contains all the elements of a traditional city—housing, retail, office space, civic buildings, parks—only now segregated and surrounded by parking.

Contrary to what one might expect, the traditional city provides better access to the countryside, and thereby a likelier basis for an active relationship between country and city. If a resident of a contemporary American city wants to visit the countryside, he must drive through miles of suburbs to get there. These might bleed away gradually over an hour or more before our urbanite finally reaches what can properly be called a rural landscape. Even there, cheap roadside development abounds. It is hard to imagine him intuiting a connection between the place he has reached and the place he lives. The distance involved impedes the connection, in reality and in the imagination. By contrast, the resident of a traditional city can often walk to the edge of town, or at least catch a bus. Immediately upon crossing a distinct edge, she is in the rural countryside. The density of the traditional city permits the preservation of more farmland and open space. If the growth of the city is well planned, new neighborhoods can be added in such a way that fingers of farmland and parkland penetrate deep into the city.

As we remarked earlier, the formal recommendations of New Urbanists are not new. Their basic principles are old, many of them so old as to obscure their origins. New Urbanism is, in one sense, just a renewal of traditions of human settlement that emerged over millennia. In another sense, though, there is much new about New Urbanism. For one thing, the technological context is new. We refer to the way the automobile has comprehensively reshaped American life. The automobile is at the root of the problem of suburban sprawl. But automobiles are too useful now to be ignored or excluded in urban design. Indeed, the right amount of automobile traffic is a key ingredient in the success of most contemporary streets. Automobiles must be accommodated within neighborhoods without compromising the quality of the pedestrian experience. This is a tricky balancing act, a relatively new challenge to urban designers. Even here, though, there are historic precedents on which to draw, namely the charming pedestrian-friendly suburbs built between the World Wars all over America.

Another aspect of New Urbanism is completely new: the legal and cultural context in which traditional urban forms must be implemented. Though the physical form of New Urbanist development closely resembles pre–World War II precedents, this form can be realized now only with great effort. Regulatory roadblocks present themselves at every turn, and we no longer have a cultural consensus on how to design the built environment. This is unfortunate and disheartening. But New Urbanists are not the only people of goodwill who face such a challenge. The challenge to agrarians is essentially similar: in order to achieve the straightforward, small-scale land-use patterns that enable humans to flourish, we must now overcome a variety of systemic obstacles and a lapse of cultural memory. This brings us to consideration of three important tools New Urbanists have developed to help them put their ideas into practice.

The prevailing pattern of human settlement, suburban sprawl, is a natural result of our existing legal, financial, and development systems. The path of least resistance today is sprawl. To build a traditional neighborhood, or even a single mixed-use building, requires considerable initiative. Though institutional structures may not support New Urbanist design, most citizens do, overwhelmingly so. In order to demonstrate this support to skeptical financiers, lawyers, and government officials, New Urbanists frequently employ what they call a "visual preference survey." This is an opinion-gathering exercise, wherein citizens are shown images that depict alternative approaches to the same street, neighborhood, or development site. A survey might contrast images of an existing commercial strip lined with big box retailers with images of the same street redeveloped as a pedestrian-friendly boulevard, lined with trees and shops. Or a series of images might show an existing, undeveloped field, the field developed as a gated condominium complex, and the field developed as a mixed-use, walkable neighborhood. Respondents are asked to rate their preferences, and their collective responses can be used to show what local residents prefer. Invariably, this exercise manifests tremendous public support for traditional urban forms.

Having established a supportive climate with a visual preference survey, New Urbanists turn to a second tool that they invented in conjunction with their first project at Seaside, Florida, the charrette. A charrette is an intensive design workshop, usually a week long, focused on a particular development, neighborhood, or city. It is sponsored by a local institution, government, or developer, and held in a public place as close to the site as possible. A team of design professionals coordinates the workshop, which brings together all the stakeholders in the project, including area residents, developers, code officials, community activists, and any other interested parties. Great pains are taken to include all potential stakeholders, so that the results will have broad public support. During the workshop, the designers come up with a development plan for the community, refining it as they go in response to stakeholder concerns. After the charrette, the sponsoring institution works to implement the plan. The broad participation, concentrated time frame, and holistic approach of the charrette make it highly effective as a tool for overcoming common regulatory roadblocks.

The proposals generated at a charrette are implemented by means of a third set of tools: codes and a regulating plan. These tools ensure not only that the necessary infrastructure is built properly, but also that individual property owners build in a manner that supports the plan. Such rules were not always necessary. Not only was there once greater consensus about how to build, there was no automobile-centered alternative.

But it is not simply a matter of revising existing zoning codes. Their form, as well as their content, must change. Chief among the reasons they are so problematic, according to New Urbanists, is the fact that they are composed almost entirely of words and numbers. Their use of visual representation is limited to zoning maps, which are little more than crude blocks of color laid over a street map. New Urbanists have discovered that by representing the elements of a design code visually, they achieve better results. These results are not only better in appearance, they are better for democratic politics. Simple requirements about the location of front porches or the heights of buildings quickly become convoluted, even indecipherable, when described in words and numbers, but drawings can establish a standard that the public can understand, endorse, and hold one another accountable to.

Points of Contact, Points of Contrast

To the reader steeped in agrarian writings, the foregoing summary will no doubt suggest similarities between the beliefs and aims of New Urbanists and those of Wendell Berry, Gene Logsdon, and others. The most obvious similarity between the two groups, perhaps, is their shared criticism of land-use patterns since World War II. Both groups complain that we use too much land, too quickly, and thereby fall into carelessness. Each group regards this tendency as having tragic consequences for human communities and the natural environment alike. Moreover, each group recommends, as a remedy, more intensive use of smaller parcels of land—that is, subtler, more diverse, more attentive use of the land, with an eye always on how to maintain it in use. What each group recommends is a renewed attention to the crafts of land shaping. Both of them observe that human settlements, traditionally organized, are improved, not degraded, by the presence of more people. Even the language of the two groups converges when they discuss this point. Thus Jane Jacobs and her followers stress the necessity of "eyes on the street," if public spaces are to feel and be inhabited, cherished, safe; and Berry and Wes Jackson talk about an appropriate "eyes-to-acres ratio," which one cannot disregard if one wants to use land discriminatingly and well. The point is, in each case, the same: that which is closely watched tends to receive better care than that which is not—provided the watchers have a long-term stake in the well-being of what they are watching.

This emphasis on land preservation has brought each group into fruitful alliances with environmental-protection groups. Like the agrarians and the New Urbanists, environmentalists care about how we treat the land. They would have us use less of it, in order to preserve more as habitat for other creatures. And they are concerned to see that the ways we use the land do as little damage to it as possible. But these alliances are restive. The source of tension is the unapologetic humanism of agrarians and New Urbanists. They are both humanistic, in the sense that each aims at the improvement of an important human practice, toward the end of helping human communities and their members flourish. In consequence, both come into conflict with the anti-humanism of certain participants in the environmental movement, those who adopt the radical and ultimately hopeless view that any detectable human influence is degradation. Both agrarians and New Urbanists propose the orderly development or cultivation of the natural environment, whether in the form of buildings, streets, and parks, or in the form of homesteads, fields, and pastures. Both regard this as a means of environmental protection, not a failure thereof.

The humanisms of agrarians and New Urbanists can also be classed together, in contradistinction to other humanisms, by the robustness and detail of their notions of human flourishing. In this, agrarianism and New Urbanism differ from, for example, the humanism of contemporary human rights doctrines. These doctrines say little more about the good life than that it presupposes food, shelter, bodily integrity, and a measure of self-determination. Agrarians, while denying none of this, go on to assert that a flourishing life standardly incorporates, among other things, interdependence with neighbors in a geographically limited, relatively self-sufficient, intergenerationally stable community; intimate knowledge of that community and its environs, particularly of the places one lives and works; close proximity, if not identity, between these places; and a measure of personal self-sufficiency through physical labor, preferably on one's own property.

With these agrarian commitments in mind, consider the view of the good life implicit in New Urbanist theory and practice. New Urbanists wish to reconnect town and city dwellers with the places they live by facilitating the development of small communities, neighborhoods, with the following characteristics, among others: pedestrian-oriented—that is to say, human-oriented—design, so that people will have occasion to interact with one another, both in the market and near home; a variety of businesses, so that the households of the neighborhood can meet most of their weekly needs without leaving the neighborhood; a consequent abundance of employment opportunities, including opportunities for self-employment, so that people can work very near home, or even at home in live-work units; a variety of housing types, so that people of all classes, at all stages of life, can have a place within the neighborhood; and neighborhood schools, parks, and civic buildings, to embody and encourage a thriving public life.

Would it be too strong to say that what the agrarian advocates in general, with special reference to rural contexts, the New Urbanist is working to realize in urban contexts? Whether by remonstrating with them or by offering structural incentives, each group, in effect, urges people to do as Berry says: "get out of your car, off your horse, and walk over the ground."[1] More important, each is working toward a common goal: a proliferation of small, relatively self-sufficient communities, populated by people with intimate knowledge of their place and their neighbors.

One further similarity between the two groups bears mention here. Both groups are unashamed to adopt and adapt traditional practices in service of their ends. They both view tradition as an organic body of generally sound judgments about good practice. They recognize that traditions change over time, the challenge being to absorb improvements while ignoring fads. Neither agrarians nor New Urbanists are opposed to progress, but each insists that progress occurs in conversation with the past. Unsurprisingly, both groups are routinely accused of nostalgia. This, however, is not a sensible objection, for why should it count against the reasonableness of a practice that it has a long history? Opposing the common prejudice against all things old, both groups seek renewal through retrieval.

We do not wish to overemphasize the similarities. There are, to be sure, differences between New Urbanists and agrarians. There is, in the first place, the obvious difference: agrarians focus, in the first instance, on rural places, whereas New Urbanists focus, in the first instance, on urban places. This difference is partly responsible, no doubt, for keeping the two groups from recognizing how much they share. But this difference is neither the only one, nor the most significant.

Arguably the most significant differences between agrarians and New

Urbanists stem from the fact that agrarianism is or involves a comprehensive political, economic, and moral standpoint, while New Urbanism is, in the first instance, a thesis about the physical structure of urban communities. Note how the agrarian account of the good life repeatedly invokes qualitative judgments about persons and social conditions. These appear also in the New Urbanist account of a good neighborhood, but there they are relegated to the status of motivating reasons. What the New Urbanist proposes, in the first instance, is a series of changes to the built environment. One could advocate New Urbanist design on purely aesthetic grounds.

By way of concluding this survey, let us note one further cluster of differences between the two groups, the vast differences they exhibit in temperament and group culture. Both agrarians and New Urbanists have, as we indicated earlier, a powerful affection for tradition. This places each group importantly out of step with contemporary mass culture. There is a difference between them, though, in the attitudes they characteristically adopt toward this displacement. New Urbanists, perhaps because they are "new," take an optimistic stance. They are highly organized and are working in an impressively coordinated manner toward their end: a revolution in settlement practices. Most of them believe that this can happen in their lifetimes. They are confident that, if bad urban design ascended through a combination of perverse incentives and market forces, good urban design can take its place through a combination of intelligent incentives and market forces. They are convinced, then, that the disagreement between them and the broader culture goes only so deep, and can be surmounted; they have a product, they believe, with greater appeal than sprawl, if only they can bring it to market.

Agrarians, by contrast, have been watching the abuse and disintegration of what they love for generations. They write, increasingly, as people whose only practical hope is an enormous catastrophe or, short of that, the proliferation of marginal groups practicing sanity in an insane, perhaps doomed world. If they have a grander hope, it is that, somehow, enough people will join these marginal groups and that they will cease to be marginal—that their influence will begin to be felt in politics, or by markets. They do not expect to live to see this result.

Learning from the Encounter

Any time two groups share an agenda to the extent that agrarians and New Urbanists do, it is worth reflecting on their differences and what can be learned from them. The first and most obvious lesson for agrarians is that the built environment has an enormous influence on the character of local communities, for good or for ill. Design matters. Specifically, traditional neighborhood design is an approach that furthers the economic and cultural health of local communities by encouraging neighborly interaction, elevating the needs of people above those of machines, and making public places lovely and inviting. Moreover, it does these things with thrift. It uses far less land than the available alternative. And it does these things by reaching back to a neglected tradition of craftsmanship. Traditional neighborhood design, to borrow a line from Berry, "is good work." Agrarians would do well to advocate it, for rural towns and cities alike.

In addition to the good this would accomplish directly, it would formalize an alliance that could serve the agrarian cause in other ways. Berry complains in one of his essays that there is no significant urban constituency for better land use. We suspect he is wrong about this, insofar as the constituency for traditional neighborhood design is significant and growing. But if his point is that these urbanites have not extended their concerns about land use to the ways land is used in the countryside, an alliance between agrarians and New Urbanists could help them to see the need for this.

In addition to this straightforward recommendation, agrarians might also make use of a range of techniques New Urbanists have developed. We wonder, first, whether agrarians might profit from more extensive use of visual representation. Visuality has not been incidental to the New Urbanist project. Every aspect of the project has depended heavily on visual representation, from the study of old neighborhoods to the design of new ones, from persuading skeptics to implementing effective codes. This is not a merely pragmatic choice; it is principled. New Urbanists argue that, notwithstanding all our manipulative commercial imagery, visuality is generally neglected in our culture, which puts too much emphasis on words and numbers. This neglect fed directly into modern zoning laws, which lack the visual sophistication needed to order a humane built environment.

To be sure, agrarians are not unaware of the importance of visuality. The texts of Berry and others are replete with sensuous, vivid descriptions of the rural landscape. But it is still worth speculating about ways that agrarians might profit from greater use of visual representation. Obviously, the regulating plans New Urbanists draw of cities and neighborhoods might find application in rural areas, where they could help communities develop specific, long-term visions of how they want to shape their common landscape. But perhaps the most promising application of visual media for agrarians would be in the recording and transmission of tradition. Since oral traditions have been mostly broken or lost, many small farmers today who sympathize with agrarian values lack the neighborly advice and community discussion necessary to sustain their practice. Visual media, especially when supplemented by perspicuous text, can be a remarkably effective communication device, surpassing unillustrated texts in terms of accessibility. In this way, magazines and well-illustrated manuals, even instructional videos, might stand as surrogate mentors to farmers who aspire to agrarian ideals. Rodale Press has done a great deal of work in this direction. But Logsdon's *Practical Skills* has been out of print for years, and nothing has taken its place. More is needed.

Agrarians might also take interest in the organizational strategies of New Urbanists. New Urbanists have been remarkably successful against the overwhelming regulatory and financial structures that generate sprawl. The first New Urbanist development, Seaside, Florida, was begun in 1981; there are now over two hundred New Urbanist developments completed or under construction around the country. Whole cities, like Milwaukee and Charleston, South Carolina, are adopting New Urbanist codes, and some states, like Maryland and Wisconsin, are tackling regional design issues.

A major factor in the New Urbanists' success is their high degree of organization, reflected more than anywhere else in the Congress for the New Urbanism (CNU). The CNU is a nonprofit organization, founded in 1992 to support local and regional efforts at implementing traditional neighborhood design. It employs several means in pursuit of this end. Most important, it provides forums for leading thinkers and practitioners to discuss and refine publicly their shared agenda. Though the conversation is often vigorous, the participants work toward common approaches and nomenclature. The achievement of a unified message has been vital to the New Urbanists' success in building a national constituency. The most important forum is the annual gathering, called a congress. The congresses provide both an occasion for the discussion of major issues and a venue for like-minded individuals to meet one another and form professional networks. A second type of forum is the task force. These more focused groups, composed of CNU members, pursue a wide range of initiatives that inform the practice of traditional neighborhood design, ranging from studies of appropriate school size to collaborative efforts with sympathetic groups such as disability rights activists, preservationists, housing advocates, environmentalists, and transportation engineers. In addition to these forums for discussion and exploration, the CNU advocates traditional neighborhood design, inside and outside the design professions. It pursues legislative reform at various levels of government. And it spreads the word through congresses, a variety of publications, speakers provided to interested groups, design competitions, and other means.

To date, there is no equivalent organization among agrarians. There are numerous and laudable local initiatives, and several organizations, like the Rural Trust, that work nationally to promote some aspect of the agrarian vision. But there is no national forum dedicated to refining the agrarian agenda, no national network by means of which agrarians can locate and assist one another, no national organization advocating the cause. Agrarians, of course, belong to a very old tradition of independent thought and action. Some will resist the suggestion that they organize themselves into anything as large and polished as the CNU. Such reluctance may be merely temperamental. But it may also stem from a deep-seated pessimism about the possibility of positive cultural change that does not begin with personal transformation.

The agrarian tradition, particularly in America, is a tradition of decentralist economics and politics. Agrarians have long held, for instance, that local communities should be more self-determining than they are, and that a broad distribution of usable property is among the most important of economic goals. They have held, as a general principle, that those most intimately familiar with a place are best situated to address its problems. All of this might appear to sit badly with the formation of a national organization. Indeed, this "think little" stance has sometimes been understood as implying the rejection of all political or economic centralization.

We think this is both a substantive political mistake and a mistaken interpretation of the agrarian tradition. Most figures in the agrarian tradition have generally held that economic and political power should be decentralized wherever decentralizing would not jeopardize the very individuals and communities to whom power would be distributed. Of course, it is a matter of judgment when this is the case; but there is consensus within the tradition about some important cases. For example, few if any agrarians have suggested that we decentralize national defense. And agrarians have not precluded the use of milder forms of state power, such as taxation, to attain goals like a wide distribution of usable property. Agrarians have been most comfortable advocating the use of state power in instances where there is a centralized threat to local communities, like a monopolistic business or an invading army. The most adequate response to such threats frequently requires centralization. Thus, national legislation that unnecessarily disadvantages small producers can and should be met with national legislative reform. Centralization, as such, is not excluded by agrarian principle.

In addition, there are two things to be said about New Urbanism that may forestall objections to a similar organization for agrarians. First, in judging the CNU as a model, it is important not to overestimate the scale of the CNU. In spite of considerable success, the CNU hardly dominates the world of development, architecture, and planning. Its two-hundred-plus developments represent a tiny percentage of the market. In order to achieve even that much, New Urbanists needed the concentrated resources of the CNU. Second, agrarians should note that the CNU emerged out of a context of academic research, debate, and prophetic writing that resembles the current context of agrarianism. Some early advocates of traditional urban design were reluctant to form a national organization. They feared that such an organization, if successful, would wrest power from local stakeholders and silence dissenting views. But contrary to their expectations, the CNU has remained remarkably open to both internal and external critique, in addition to furthering research into urbanism.

Does the situation agrarians now confront call for a centralized response? Local communities are everywhere threatened by policies set at the state and national levels, and by the expansionist behavior of many corporations. A unified, national voice is needed, at the very least, to offer legislative alternatives and arguments. This, in turn, requires a forum for agrarians to discuss their differences and move toward consensus, an organization that could disseminate the views of agrarians to a wider audience and offer national support and publicity to local and regional initiatives.

Whether or not agrarians agree to organize nationally, they can certainly make use of one New Urbanist device: the design charrette. What a charrette is, among other things, is an approach to problem solving, one designed to address complex issues of public concern within a relatively confined geographical area. It has many strengths. Its holistic approach and restricted time frame tend to bring opposed interest groups together and force them into serious reflection on the possibilities of compromise. One of the charrette's greatest strengths, we feel, is its power to simultaneously guide and respond to public sentiment, with the result that fundamentally healthy patterns can be implemented with full and informed local consent, rather than by fiat from without; moreover, these patterns can be implemented in ways that respond subtly to local concerns and conditions, rather than as context-independent abstractions.

An equally great strength is the way the charrette places professional expertise at the disposal of communities looking to act on their own behalf. Here the credentialed expertise and creativity of the design team is crucially important. These qualities enable them to address with authority objections from (outside) corporate and regulatory bodies, objections that all too often stop local initiatives in their tracks. The result, when all goes well, is a comprehensive plan for the neighborhood, approved in concept by all parties and therefore well on its way to implementation.

The charrette model thus represents one possible answer to a question that has long vexed agrarians: How does one address complex community problems in an effective way that begins to embody a decentralist politics? Berry has repeatedly warned about the dangers of relying on large-scale, state-level solutions to complex community problems. He urges us to focus first and foremost on improving individual practice, or "home economics." This emphasis on individual and family responsibility is perfectly appropriate. But is there nothing more to be done, without reverting to the use of state power? Is there no way for communities to begin to act on their own behalf? The charrette model offers a possible third way, one that complements individual and family responsibility, while avoiding the use of state power. Its uses can be expanded and adapted in various ways to meet particular, local needs, such as the need for deinstitutionalized forms of elder care or the need to develop value-adding industries for local products. Of course, the charrette is not an omnicompetent tool. Nothing is. But it is versatile. And contemplate what might be accomplished, community by community, if agrarians with legal, economic, and other expertise were to offer their services to community groups that wished to tackle these types of challenges.

Closing Reflections

So far, we have focused mainly on what agrarians can learn from New Urbanists. But there is no reason to think this exchange should go only one way. What might New Urbanists learn from agrarians? First, there is the simple point that New Urbanists have much to learn about farmland and rural places generally. This is not to say that the New Urbanists are unconscious of the value of these places. Quite the contrary. The *Charter of the New Urbanism* asks: "In this era of modern agriculture, do efforts to save farmland amount to little more than a sentimental gesture? The answer is that saving farmland and other agricultural land remains crucial to the health of metropolitan communities" (29). The reasons put forth range from the preservation of open space, to the protection of water supplies, to the availability of fresh produce. While this defense of farmland goes beyond a merely aesthetic response, it is undeniably written from the perspective of a city dweller. Given the reciprocal relationship that both groups wish to see between cities and rural places, New Urbanists need to deepen and nuance their understanding of the rural—or, as they sometimes refer to it, the "agricultural hinterland." Agrarians have a vision of farms and farming communities that is positive, not merely derivative from the city. This vision could inform and complement the New Urbanist model, contributing to a broader vision of how humans can best inhabit the land.

But there is a deeper, more substantial challenge that an encounter with agrarians would pose for New Urbanists: namely, to come to terms with the radical economic and cultural changes implied in a thoroughgoing return to traditional neighborhood design. Agrarians, we think, are more cognizant of what they are asking people to surrender—indeed, *that* they are asking them to surrender something. Contrary to popular belief, it is not possible to have it all: to pay less for our food than it is worth and to preserve small-scale agriculture at the same time. Agrarians have called us to confront this choice and others like it. They have argued that what we would surrender to preserve local communities is comparatively unimportant. But they have not evaded the implications of their work: that Americans should pay more for the food they buy, and grow some for themselves; that people should be less "mobile"; that smart people should spend their intelligence on small, unglamorous problems in small, unglamorous places; that we should be slow to adopt technological innovations; that people should house their aging parents rather than commit them to the care of institutions.

That New Urbanists have dealt satisfactorily with such issues is unclear. Again, the *Charter:* "We recognize that physical solutions by themselves will not solve social and economic problems, but neither can economic vitality, community stability, and environmental health be sustained without a coherent and supportive physical framework" (v). The New Urbanists have vigorously pursued the "physical framework," leaving the other dimensions of these problems unaddressed. New Urbanists have certainly succeeded in building communities with tremendous appeal. But people also find Wal-Mart appealing; they just prefer not to live next door to it. People undoubtedly feel the pull of neighborliness. But is this affection strong enough to make them abandon the security and cozy homogeneity of gated communities? It is not unheard of, after all, for people to be attracted to the good and choose against it all the same. Addicts display this sort of behavior. Indeed, as Berry suggests in several places, we have become an addicted culture. And there is much evidence to support his view.

What New Urbanists must confront is the likelihood that communal and cultural health, like bodily health, like any virtue, requires ascetic discipline.[2] Such discipline is not, of course, incompatible with traditional neighborhood design. Indeed, good physical design helps cultivate the virtues. We do not, therefore, retract the recommendations we have made. But recognition of the relationship between communal health and ascetic discipline does affect the way one thinks about, and works at, community building. It draws one beyond the marketing of a useful product, and toward comprehensive social and political reflection. It impresses the thought that one may not live to see the necessary work completed. It enforces humility. And it points up the importance of what Berry aspires to in virtually everything he writes: persuading people, one by one, to discipline themselves.

Recommended Reading

- Leccese, Michael and Kathleen McCormick, eds. *Charter of the New Urbanism*. New York: McGraw-Hill, 2000. The charter is a remarkably clear description of what the New Urbanism is all about. Twenty-seven succinct and well-illustrated essays by twenty-seven leading New Urbanists deal systematically with every scale of concern.
- Bess, Philip. "Virtuous Reality: Aristotle, Critical Realism and the Reconstruction of Architectural and Urban Theory." *The Classicist,* no. 3 (1996). Philip Bess, a practicing New Urbanist, offers a sophisticated internal critique of New Urbanism and the cultural context that might be necessary to implement its vision.

- Duany, Andres, et. al., *Suburban Nation: The Rise of Sprawl and the Decline of the American Dream*. New York: North Point Press, 2000. Written by the leading practitioners of New Urbanism, this is the clearest exposition of New Urbanism to date.
- Kunstler, James Howard. *Home from Nowhere*. New York: Touchstone, 1998. This delightful, acerbic account of the problems of sprawl also describes the New Urbanists' solutions.
- Krier, Léon. *Architecture: Choice or Fate*. Windsor, Great Britain: Andreas Papadakis Publisher, 1998. The text is good, but the witty and engaging polemical cartoons are brilliant. They make clear the basic patterns of urbanism.
- Jacobs, Jane. *The Death and Life of Great American Cities*. 1961. Reprint, New York: The Modern Library, 1993. It took a nonprofessional to first recognize the devastation that the architecture and planning professions had wrought on great American cities. This influential, well-written wake-up call is now a classic.

Notes

1. Wendell Berry, "Out of Your Car, Off Your Horse," in *Sex, Economy, Freedom, & Community* (New York: Pantheon, 1992), 20.

2. We are indebted, on this point as on many others, to Philip Bess. We regret that this essay's restricted length prevents us from doing justice to this idea.

12

NEW AGRARIANS

Local Innovators

Susan Witt

A recent rereading of Wendell Berry's *The Unsettling of America* occasioned a reflection on the role and promise of agrarianism in these changing times. The events of September 2001 have helped end the lingering enchantment with the monoculture of the global economy. The consequences of our dependence on the products of a relatively few international corporations are now more visible, including the concentration of ownership of the means of production. The consumption patterns of a small number of countries account for most of the depletion of the earth's natural resources. Environmental degradation occurs out of sight of the end consumer, and so the responsibility for restoration goes unheeded.

Instead of addressing the conditions of poverty, the global economy has only exaggerated the intolerable discrepancies in income and distribution of goods around the world. The poor are even further impoverished when, following the promise of available jobs, they are enticed into the enclosed compounds of corporate factories with the accompanying loss of the social fabric that has nourished them. The cultural life and traditional skills of the village are left behind—skills that if cultivated could provide the basis for a richer local economy in which the production of the basic necessities of food, clothing, energy, and shelter could better be met. These glaring discrepancies breed resentments and hostilities, creating volatile conditions that are used to justify increasingly powerful and dangerous forces of repression.

But quietly and surely around the world, new attention is being focused on the renewal of village economies. In village after village, leaders are appearing whose roots run deep in their local community. They do not need outside consultants to show them the natural riches and human skills available to shape new patterns of local production and local trade. They are using their imagination to craft new local institutions to support this renewal. It is these villagers, both rural and urban, who are the "new agrarians," creating the basis for a new peace, while champions of the global economy are risking the lives of us all by fostering the conditions for a new war.

What are the characteristics of these new agrarians and their village economies? In his book *Why the Village Movement?* Gandhian economist J.C. Kumarappa addresses the women of the villages with these thoughts: You, my sisters, perhaps it is your husbands who earn the money for your family, but it is you who are determining how that money is spent; in so doing you are deciding the fate of your village. You may choose to buy the beautiful silk made in France or Belgium, or you may choose the khadi cloth made by your sister and your neighbor. When you choose the khadi cloth, you are investing in more than cloth—you are investing in your neighbor, her children, and your village. As you watch the children walking to school in the morning, fed by the earnings of their mother, you realize that you and they are woven together through the cloth. You and your village are richer in proportion to the number of stories that unite you.

Across North America, in region after region, citizens are banding together in their role as consumers to work with producers, sharing the risk of production costs in order to help shape the kind of vital local economy that incorporates social and ecological objectives. I believe that the future of agrarianism lies with these regionally based producer/consumer associations.

In 1986, in my own Jug End Road neighborhood of the Berkshires region of western Massachusetts, Robyn Van En founded the first community supported agriculture (CSA) project in the United States at her Indian Line Farm. In a CSA, consumers guarantee the yearly production costs of the farmer through a shareholder fee. Working in collaboration with shareholders, the farmer determines an annual operating budget. Ideally, the budget is then divided by the number of shareholders to determine the cost per share. CSA members pay in advance so that funds are available to the farmer during the growing season. In return they receive a weekly share of the harvest and the security of a local source of organically raised vegetables. Because of Van En's initiative there are now over one thousand CSA farms around the country.

CSAs provide an excellent model for consumers sharing the risk of yearly production costs with the farmer. Yet the question remains: How can young farmers gain affordable access to land in the first place? Again Indian Line Farm provides a model. When Van En's farm came up for sale following her untimely death in 1997, the sale price was too high for entering farmers. A farm income alone could not carry mortgage payments and still maintain responsible farm practices.

The rising cost of the land put the purchase price out of reach, a typical problem in regions close to urban areas or deemed valuable for vacation homes. The market value of the land reflects the demand for house sites, frequently second-home sites, rather than the social benefit of maintaining a local farm. High purchase costs of the land and the pressure of tax and mortgage payments on that purchase can drive a farmer to employ unwise farm practices and production methods beyond what is ecologically suitable for the land. If the citizens of the southern Berkshires wanted Indian Line to remain an active farm producing vegetables for local sale, they would have to partner with the farmer to purchase the farm.

The community, working through the Community Land Trust in the Southern Berkshires and The Berkshire Taconic Landscape Program of The Nature Conservancy, made a one-time donation to purchase the land. The Community Land Trust holds title to the land, and The Nature Conservancy holds a conservation restriction. This arrangement has enabled two young farmers, Elizabeth Keen and Alexander Thorp, to purchase the buildings and enter into a ninety-nine-year lease on the land, the use of which is determined by a detailed land-use plan.

The individuals who donated to the project needed the incentive of knowing that the Community Land Trust and The Nature Conservancy would not be coming back next year to the donors to refinance the same farm—that this one donation would keep the farm actively farmed and affordable for future farmers. Ownership, after all, is only a bundle of rights. It was simply a question of what rights the donors, working through the two nonprofits, wanted to retain in return for their role in purchasing the land; put another way, how many rights would the farmers require to preserve the incentive to farm with all their heart and strength?

The Nature Conservancy used the legal tool of a conservation restriction to protect the ecological quality of the land for future generations. The Community Land Trust used the legal tool of a lease to ensure that community objectives are maintained. The lease has the following requirements:

- The buildings are to remain occupied by those leasing the land. They cannot become rental property or vacation homes for city people.
- The land must in fact be farmed. The lease requires a minimum yearly commercial crop production over and above household use; however, the lease does not specify what kind of crops should be grown. This is the private affair of the farmers, based on their evaluation of local markets.
- At resale the buildings must remain affordable to the next farmer. The lease requires the leaseholder to offer all buildings and other improvements back to the Community Land Trust in the Southern Berkshires for resale. The price can be no more than the current replacement cost of the buildings, adjusted for deterioration.
- The farmer is to employ organic practices and meet the conditions of a land-use plan developed to respect the specific ecology of the site.

All other ownership rights belong to the farmer, including the ability to pass on the farm to heirs through the transfer of the lease.

The lease arrangement does not guarantee that the farmer will farm well. Such skills are cultural and acquired over many years. However, by taking away the burden of land debt, the community land trust gives farmers the opportunity to ply their craft under more favorable circumstances. As land prices continue to rise in regions surrounding cities, it will be ever more important for citizens to utilize such enabling methods to ensure a local food supply.

The basic achievement of the community land trust legal documentation is to separate the value of the land from the value of buildings and other improvements on the land (such as fences, soil fertility, perennial stock). Land, a limited natural resource, is removed from the market and held in trust by the democratically structured, regional nonprofit. The value created from labor applied to the land (agricultural crops, buildings, etc.) is securely the private equity of the person creating the value (the farmer) and is exchangeable in the marketplace.

The private, nonprofit community land trust is thus a flexible tool for a community to determine its own goals for land use and distribution. As a result of the success of the Indian Line Farm model, several conservation land trusts in the Berkshires and nearby Connecticut are in the process of initiating their own community land trusts in order to have a complex of legal tools at hand with which to protect not only open farmland but the affordability of homes and farm buildings for the future farmers working the land.

Affordable access to land for homes, small businesses, and local production is key to vibrant local economies. During the next two decades we will see community activists ever more engaged in land reform issues. One important model for this reform is the work of Vinoba Bhave, the trusted associate of Gandhi. In the 1950s Vinoba walked from village to village in India seeking a way to alleviate the widening social disparities he saw. Wherever he went, crowds gathered to listen to their beloved spiritual leader. Vinoba asked boldly, "Those of you with more land than you need, would you not give your excess to my brothers and sisters who have no land to build their homes or cultivate their crops?" Moved by the man and his appeal, wealthy villagers deeded their land to Vinoba so that he could reconvey it to the landless.

In this way the *Bhoodan* (Land Gift) movement was born. But soon

Vinoba saw that the poor, who did not have the money to buy tools to work the land and seeds to sow it, simply sold the land back to the wealthy. They then wandered into the even greater poverty of the cities. As a result, Vinoba changed the Bhoodan movement to the *Gramdan* (Village Gift) movement. The land was given to the village, and villagers were given use rights. They were not tempted to trade the land for quick money. If they left the land, it was redistributed to those who could use it.

The Gramdan movement was introduced to this country by Robert Swann, founding president of the E.F. Schumacher Society. A carpenter by trade, he used his skills during the civil rights movement of the 1960s to help rebuild bombed churches in the rural South. There he met Slater King, the cousin of Martin Luther King, and other civil rights leaders. From them he learned that African Americans were being prevented from gaining access to land. It was a pressing problem that fueled the civil unrest.

Swann worked with Slater King to adopt Vinoba's model. New Communities in rural Georgia was the first community land trust in North America, formed to provide access to land for African American farmers. There are now more than one hundred community land trusts around the United States in both urban and rural communities—started by church groups, community development corporations, concerned citizens, conservation groups, and state agencies. As each new initiative breaks ground in its region, develops a land-use plan, provides a lease, and secures a mortgage, the understanding that there are alternatives to private land tenure grows.

The agrarian writers of the 1930s, as represented by the contributors to *I'll Take My Stand: The South and the Agrarian Tradition*, considered private land ownership a cornerstone of their policies. They wished for a society characterized by a democratic distribution of usable property. But the exchange of land, a limited resource, in the marketplace has led inevitably to speculation in land and the resulting accumulation of large holdings in fewer and fewer hands. I would argue that the community land trust model offers greater opportunity to achieve the agrarian vision of small landholders beholden only to the soil and the community surrounding it. A community land trust is a nongovernmental, citizen-based approach to achieving this vision. It uses the resources and good will of the local community to gather parcels of land held in private ownership and transfer them into public trust. Applying these village-based economic tools successfully demands our finest skills as human beings. It requires imagination, local knowledge, entrepreneurship, a long-term commitment to place, and strong working relationships with others—all characteristics of new agrarians.

In many ways it is easy for us to become passive consumers of the products of global corporations, in part because the manufacturing process is invisible. Creating vital local economies requires our active engagement and tests our convictions. A community land trust is not a government tool for reallocating land. It is a citizen-based tool. But a community land trust is only as effective as the will of a region's residents to become involved in the process. They must look beyond home economics to community economics and understand that the health of the community is part and parcel of our own health and vibrancy—a vision Wendell Berry so beautifully articulates in *The Unsettling of America*.

Broad-based democratic access to land is only one aspect of an agrarian economy. A credit system that favors small locally owned businesses is also essential. The Self-Help Association for a Regional Economy (SHARE) in my home region of the Berkshires provides a model for consumers to share the risk in the start-up costs of small businesses. The objective of SHARE is to make productive loans to people who are unable to secure normal bank financing but who have the kind of small, locally owned enterprises that produce quality goods and services for local consumption. SHARE members open savings accounts at a local participating bank, and these accounts are used by SHARE to collateralize loans. This kind of lending requires that the community separate the functions of banking: the bank makes the loans and handles the accounting, but the lending decisions—based on a set of social, ecological, and financial criteria established by SHARE—are made by the community of depositors.

SHARE has collateralized loans to a goat-cheese producer for the stainless steel equipment in the milking room, to a home knitter for a bulk supply of yarn, to the owner of a working draft-horse team for materials for a barn, to a kite maker for a quantity of waterproof fabric, and to a music teacher for a piano. The payback record on SHARE-collateralized loans has been 100 percent, both because of their scale and because of community support for the loan recipients. SHARE members help maintain this perfect record by recommending these small businesses to their friends.

The SHARE loan-collateralization program is simple to operate and easily copied. Similar programs have started around the United States, using the model created in the Berkshires. It is the "grandmother principle" that has made SHARE a success. When people without credit histories decide to go into business, they frequently turn to a family member, such as a grandmother, for help. Instead of lending directly, the grandmother might offer a savings account as collateral for a bank loan. The SHARE program simply extends "the circle of grandmothers," creating a family based on place.

In Bangladesh, Muhammad Yunus established the Grameen Bank, a widely replicated model for making small loans. Borrowers come as a group to the bank, agreeing to share financial responsibility. The bank makes only one loan at a time to a member of the group. The unwritten social contract among neighbors, rather than a formal bank contract, helps ensure the loan's repayment. Both SHARE and the Grameen Bank depend on the availability of savings deposits—the excess funds of others—to make their loans. This dependence is a limitation. In her book *Cities and the Wealth of Nations*, the regional planner Jane Jacobs offers another approach: she describes local currency as an elegant tool for providing credit for regional businesses.

Jacobs views the economy of a region as a living entity in the process of expanding and contracting, with a local currency as the appropriate regulator of this ebbing and flowing. She advocates the creation of import-replacement businesses so that the goods and services consumed in that region are produced there. Local currencies can be placed in circulation through the making of productive loans. Productive loans are those resulting in new goods available in the economy in excess of the value of the loan itself, such as loans to a farmer for seeds in the spring that generate a bountiful crop of fall vegetables. The interest rate on loans made through the issue of local currency can be as little as 0 percent because the issuing community is not paying interest to a group of depositors such as in the SHARE program. Reducing financing charges encourages the development of small local manufacturing enterprises or renewable-energy generating plants that currently are not economically competitive. Eventually a community-based organization issuing local currency could untie its currency from the federal dollar, establishing a local backing such as cordwood or a basket of commodities—corn, soybeans, and wheat, for instance. Then currency would retain a constant local value related to a natural resource and would make visible once again the connection between the health of a local economy and the health of the land.

Currently twenty-one communities in the United States and Canada are experimenting with local issue of currency, and the movement is growing as so-called developing countries learn that they do not need to borrow from the International Monetary Fund to generate local businesses. They are freeing themselves from the global economy to focus on their local economies.

We can start the process in very small ways. Deli Dollars, issued in Great Barrington, Massachusetts, in 1989, were a way for customers of a local restaurant to support a popular business as it prepared to move to a new location. Consumers purchased Deli Dollars for eight dollars each and then, after the move was complete, cashed them in for ten dollars' worth of sandwiches and coffee. The Deli Dollars were dated over a year's time to spread out the redemption period. This local scrip, which was transferable, turned up around town, including in the collection plate of the Congregational Church because the parishioners knew that Pastor Van ate breakfast at the Deli.

The concept of these local economic tools is simple: by sharing in the risk of production and acting creatively, consumers can ensure that local businesses remain and thrive in their communities.

As we work in our own communities to bring about healthy regional economies, we are not working as isolationists, securing only our own future and our families' futures, but rather we are working in solidarity with villagers around the world who are seeking ways to revitalize their own economies. It is important to share the stories of our successes. A vigorous local economy for the Buryats in the Olkhon region on the western shore of Lake Baikal in Siberia, for example, will certainly look different from a more self-sufficient Kentucky bluegrass economy, but it will grow out of the same love of place and community.

The publication of *The Unsettling of America* awoke us to the crisis in our rural communities: Wendell Berry describes how the loss of a way of life rooted in agricultural traditions leads inevitably to cultural loss, multidimensional in its effect. Yet *The Unsettling of America* is more than a series of astute observations on the loss of rural life; it is at the same time a statement of active love for the observed. This is characteristic of all of Wendell Berry's work and defines his greatness: embedded in his observations is the moral imperative to defend what is being lost. *The Unsettling of America* leads us to seek an authentic resettlement and re-inhabitation of the place we call *home*.

In *The Unsettling of America* and the stories, essays, and poems that followed it, Wendell Berry speaks from his particular experiences with the people, land, and community of Henry County, Kentucky. His writing helped spark a new generation of writers who—instead of leaving their homes to find their literary voices as expatriates in European capitals—returned to their roots, their birth cultures, to explore the process of repatriation. In telling well the story of each particular community, this new generation of writers is helping to give voice to those hundred thousand villagers in thousands of villages around the world. It is these villagers, both rural and urban, who are defining a new agrarianism as they learn through trial and error to shape their local economies and communities. In this work they are building the basis for a new peace. Let us celebrate them and join with them as responsible citizens working creatively with local producers to fashion the future of our own neighborhoods and regions.

13

The Legal and Legislative Front

The Fight against Industrial Agriculture

Hank Graddy

Over twenty-five years ago Earl Butz gave farmers the advice to "get big or get out," advice that was part of the outrage inciting Wendell Berry to write *The Unsettling of America.*[1] Today, agrarians who argue against industrial, centralized, corporate-controlled food production in favor of sustainable agriculture question if that advice was (and is) a description of the inevitable course of history. Does the power exist not only to resist the Earl Butz course, but to defeat it? Can the fight against the army of industrial agricultural be won? If so, how?

Private, personal action is very worthwhile, even good for the soul; unfortunately, in most cases it is good for only one soul. As chair of the Sierra Club's CAFO (Concentrated Animal Feeding Operation)/Clean Water Campaign Committee, however, I recommend organized action, action that involves two or more people, and thus has the elements of a campaign: a series of organized or planned steps taken to accomplish an objective. The future of agrarianism is a function of the degree to which its supporters can turn conceptual support into organized action—into a campaign. Those considering a campaign should consider at least three important questions: Is there an actual campaign to join or use as a model? Is the campaign well planned? Is the campaign hopeless?

I offer as one example a particular Sierra Club campaign against industrial meat production. The campaign, submitted to others to think about and reexamined by the committee, is well planned. I believe this campaign will succeed—that it is not hopeless—because we have experienced success, and as we have experienced success we have seen the weakness of the enemy. While we see the enemy as hugely powerful at present, we also see the enemy as doomed.

The Sierra Club and Agrarianism

Many people might be surprised to find agrarians and the Sierra Club on the same side of an issue. Started in 1892 by John Muir, the Sierra Club is an urban and suburban organization. Its advocacy for wilderness, which has always been one of its core purposes, has handicapped the club's ability to work with farmers because, by definition, wilderness lands are not to be harvested. Farmers fear that the Sierra Club's preference for "humanless" lands reflects an organizational hostility toward all land managed or owned by the private sector. The Farm Bureau and other confederates of industrial agriculture never miss an opportunity to fuel these fears with misrepresentations designed to cause farmers to distrust the Sierra Club and other environmentalists. Unfortunately, the Sierra Club has also acted in ways that fuel the fear and distrust, perhaps beginning with John Muir's reference to sheep as "hooved locusts."[2]

Fortunately, the Sierra Club has members who are farmers. Bob Warrick, a farmer from Nebraska and former co-chair with me of the club's Agricultural Committee, helped the Sierra Club plan to influence the 1985 Farm Bill, resulting in landmark conservation features such as conservation compliance and the conservation reserve program. Also, the Sierra Club has had an agricultural policy since 1976; it expresses a preference for appropriately scaled family farm agricultural operations and urges consumers to eat lower on the food chain by eating less meat.[3] More recently, the Sierra Club has focused a great deal of attention on intertwined agricultural and environmental issues: CAFOs and clean water.

THE CLEAN WATER ACT—MISSING, INACTION

The Clean Water Act, 33 USC section 1251, et seq., generally exempts agricultural operations from any specific requirements. That law divides the world of water pollution into two categories: "point source pollution" and "nonpoint source pollution."[4] Point sources of pollution have clear requirements. They are illegal unless people have applied for and obtained a permit to discharge, called a "national pollutant discharge elimination's system" or NPDES permit (in Kentucky the permit is a KPDES permit).[5]

The act defines "nonpoint source pollution" as everything else that causes water pollution.[6] Nonpoint sources of pollution have no requirements under the law until or unless these sources impair water quality as stipulated under the Clean Water Act's total maximum daily load (TMDL) requirement.[7]

Thirty years of experience with point source pollution throughout the United States have resulted in a well-established regulatory system. However, one very conspicuous gap in the regulation of point sources remains. Tucked innocently in the above definition of "point source" are the following four words: "concentrated animal feeding operations" (CAFOs).

Congress acted purposely to regulate CAFOs, just as Congress intended to regulate municipal, publicly owned treatment works and private industrial and residential wastewater treatment plants. The legislative history of the Clean Water Act includes the following 1972 quote from Senator Robert Dole: "Animal and poultry waste, until recent years, has not been considered a major pollutant. . . . The picture has changed dramatically, however, as the development of intensive livestock and poultry production on feedlots and in modern buildings has created massive concentrations of manure in small areas. The recycling capacity of the soil and plant cover has been surpassed. . . . The present situation and the outlook for future developments in livestock and poultry production show that waste management systems are required to prevent waste generated in concentrated production areas from causing serious harm to surface and ground waters."[8]

Unfortunately, Congress's clear intent to regulate concentrated animal feeding operations got lost in the implementation process. U.S. EPA regulations for CAFOs have had the practical effect of completely ignoring them. The regulation said, in effect, that an animal facility keeping under roof over 1000 "animal units" (about 1000 beef cows, 750 dairy cows, or 100,000 chickens) must design its waste handling features so as not to have any discharge except during a very heavy rain (twenty-five-year, twenty-four-hour rain event); such a design relieved the facility of any other operating, monitoring, inspecting, or reporting requirements. Essentially the EPA regulations gave great freedom to large, confined animal facilities that weren't planning to pollute.[9]

Even this regulation was more than most states actually implemented. Most states simply ignored this part of the Clean Water Act. Poultry was excused from any compliance requirements based upon its use of dry litter. The environmental community and most citizens also ignored this Clean Water Act requirement for almost twenty years.

In the mid-1990s after the concentration and industrialization of poultry production was virtually complete, and as swine production was rapidly moving to emulate poultry, Sierra Club members in the Midwest began to hear and to smell the impacts of the recently constructed poultry and swine operations. In this period large-scale animal facilities severely impacted certain streams and lakes in Missouri (flowing from Arkansas), Oklahoma, Iowa, and Illinois.[10] Rural communities where new CAFOs were proposed asked the Sierra Club for help. Examining the Clean Water Act point source permitting process, we discovered a virtually nonexistent program, a quick, rubber stamp process excluding public participation.[11] By this time a few cases involving Clean Water Act compliance arose.[12]

The Natural Resources Defense Council (NRDC) responded to the EPA and state failure to implement this aspect of the Clean Water Act by suing the EPA. As part of the settlement agreement of NRDC vs. EPA,[13] the EPA agreed to update the CAFO regulations, to propose draft regulations in ten years, and to finalize them two years later. On December 15, 2000, the EPA complied with this requirement by publishing proposed CAFO regulations, and on December 15, 2002, EPA published the final CAFO rules.[14]

Cagle's-Keystone Comes to Kentucky

In Kentucky the legal challenges to factory farms commenced with litigation by the Sierra Club and others concerning the Cagle's-Keystone poultry-processing facility in Clinton County, Kentucky. This litigation included a National Environmental Policy Act (NEPA)[15] challenge in response to the decision by the USDA to award money to the city of Albany to expand its municipal water treatment facility to provide water to the Cagle's-Keystone chicken-processing plant. It also included a challenge to the so-called "no-discharge permit" that the Kentucky Division of Water issued to the Cagle's-Keystone facility. In one sense, both of these suits were unsuccessful. The Sixth Circuit Court of Appeals dismissed the NEPA challenge as being moot because the plant was built after the District Court found that the USDA had properly performed its duties under NEPA. The permit challenge relied upon testimony predicting that wastewater sprayed on the "hay farm" located in a karst area of sinkholes, caverns, and underground streams would not prevent water quality problems downstream in Lake Cumberland.[16] The hearing officer accepted Kentucky Division of Water testimony that water quality problems would be addressed if they developed.

In another sense, these challenges yielded several positive results. The NEPA challenge required the USDA to complete an environmental impact statement (EIS). The Sierra Club argued that the EIS was a superficial analysis of environmental impacts, but the challenge helped raise public awareness about the foreseeable environmental consequences of this industry, something that had not happened when the three previous chicken-processing plants in Kentucky were built. The first serious legal challenge to industrial agriculture in Kentucky, the NEPA challenge helped frame the issues and mobilize the citizen response.

The "no-discharge" permit challenge failed to stop the permit, but the prediction that the "hay farm" was not adequate to handle the wastewater came true. Cagle's-Keystone has now purchased another farm to provide additional capacity to land-apply the water (one of the remedies sought by Sierra Club), and it appears that Cagle's-Keystone is now planning to hook into the new sewage treatment plant that Albany is building, exactly what Sierra Club sought in the permit challenge.[17] The promised jobs and economic development have failed to materialize (most of the jobs are held by migrant workers). In the November 2002 election, the county judge-executive and the fiscal court that had been so eager for Cagle's-Keystone to come to Clinton County were both defeated. An adjoining county, Cumberland County, enacted the strongest local ordinance in Kentucky to prevent adverse impacts from chicken houses.[18]

The Sierra Club National CAFO/Clean Water Campaign

The well-documented action by Cagle's-Keystone to locate a chicken-processing plant upstream from Lake Cumberland helped provide the Sierra Club with momentum to develop a national campaign to fight CAFOs. In 1997 Sierra Club activists formed a committee to address CAFO issues, and I agreed to chair that committee. The CAFO Working Group and Sierra Club staff person, Katherine Holmann, created the "Sierra Club Pig Map" to show that the CAFO problem was more than an isolated or regional issue, that it was a national and international environmental problem.

In January 1999 Sierra Club water quality activists gathered at Coolfont, West Virginia, to create the Sierra Club Water Quality Campaign Plan. We recognized that the Clean Water Act's failure to address nonpoint sources of pollution was the most serious barrier to solving poor water quality problems. We also recognized that as agriculture was becoming more concentrated and industrialized, it was becoming a more conspicuous water quality problem, just as Senator Dole had predicted in 1972. The Sierra Club decided to focus the Clean Water Campaign on CAFOs to begin to address the most neglected aspects of Clean Water Act implementation. We anticipated that if we were able to compel CAFOs to comply with Clean Water Act requirements, we would be able to improve the rest of agriculture's water quality performance.

Our initial campaign plan included three components:

1. Stop CAFOs. Where CAFOs were already in place, the campaign sought to end all of the adverse environmental impacts from the facility.

Wherever CAFOs were proposed, the campaign sought to prevent them from being built. Period.

2. Monitor water quality.
3. Educate consumers.

We have asked ourselves if it is possible to "fix" a CAFO, to prevent it from causing environmental harm. Most of us believe that these facilities cannot meet such a test. However in the interest of fairness we will continue to consider the possibility that large industrial animal facilities and processing plants can be operated in a manner that does not damage the environment. As our campaign extends our focus to include the humane treatment of animals, we may reach the conclusion that there is no possibility that any CAFO can meet appropriate standards of environmental protection, that concentrated animal feeding operations that cage animals for life cannot be "fixed," and therefore must be eliminated.

Whether or not that policy is made formal, we seek to make these facilities pay all of the costs of environmental protection. These industrial facilities have been able to externalize—make someone else pay—the costs of environmental injury and cleanup. We seek to change that practice: to make these companies pay the full costs of their environmental injury. We anticipate this requirement will make them sufficiently unprofitable that the market will do what the law may not be able to do. This part of our campaign relies on local, state, and federal legislation and regulation, and upon litigation.[19]

The part of the campaign that seeks to stop new CAFOs from being built represents the Sierra Club doing its best work: helping people who are desperate to protect their homes and their communities. Those who promote industrial animal-processing facilities typically seek out economically depressed areas that lack planning and zoning or other local controls. Companies look for the places of least resistance, places where they do not expect to confront the Sierra Club. When we are asked to help a targeted community, we send Sierra Club staff members to help them organize their opposition. When the company claims that it will use "state-of-the-art technology" to prevent environmental problems, and that it will bring good jobs and economic development, we provide the real track record for that company. We provide the community with our "toolkit"—including our newest tool, the *Rapsheet on Animal Factories.* We have been successful, and we are becoming more successful, as the record of these companies becomes more widely known.[20]

The second component of the campaign is increasing the number of qualified citizens who are trained and equipped to monitor water quality. This part of the campaign supplements efforts by state agencies to monitor water quality and provides scientific evidence that helps strengthen the case for increased protection of water quality from agricultural operations. It also provides a way for those not living under the threat of a CAFO to help the campaign.

Our goal includes identifying water quality problems that are not related to agriculture. This part of our campaign, now a separate part of the Sierra Club water quality work, is called the Sierra Water Sentinel Program; three of the eight current Sentinel programs monitor waters that may be impacted by industrial agriculture.[21]

Third, we seek to educate consumers. We recognized at Coolfont that one problem with our campaign is that people ask where our food will come from if we stop CAFOs. We recognized the need to identify an alternative source; the logical place to look for the alternative was the food production system as it existed before industrialization. We also recognized that industrial food producers had succeeded in disconnecting the consumer from the producer, probably because industrial agriculture recognized that if the consumer actually knew how her food got to her plate, she would be horrified. We recognized that educating consumers to restore the connection between production and consumption was an essential part of our campaign.

Sierra Club members began the education process by seeking to educate themselves. Our self-education included the effort to insure that the meals served at the 2002 "Future of Agrarianism" conference in Georgetown, Kentucky, were raised and processed, as much as possible, by Kentucky farmers, to insure that these farmers were paid fairly for their produce, and that they were recognized, thus making it more likely that consumers would reconnect with producers.

Our consumer education campaign focuses primarily on the use of antibiotics in the industrialized food production system. We are exploring how best to educate urban consumers about the consequences of antibiotic overuse and how to insure their access to antibiotic-free food.[22]

SERFS ON THEIR OWN FARMS

One aspect of our fight deserves special attention. The term "integrator liability" refers to one of the most appalling aspects of industrial food production. Poultry production is the most industrialized sector of agriculture, with swine production rapidly catching up. The current structure of the poultry production system, as designed by the major poultry companies, divides the capital investment and risk into two separate worlds. The chicken-processing company itself, such as Tyson, Perdue, Cagle's or Seaboard, is called the "integrator"—a term derived from the industry's "vertical integration," by which a company controls and owns several or all of the steps of production from grain and egg to kitchen table. The prevailing wisdom among industrial food proponents is that some vertical integration is better than none, and that complete vertical integration is better than partial integration.

About a dozen companies, or "integrators," control over 95 percent of U.S. poultry production. Their system requires about half of the capital investment needed for this industry and takes less than half the risk of failure. By concentrating poultry processing into the hands of a dozen integrators, these companies have been able to contract with farmers who mortgage their land to finance chicken house construction. This technique shifts about half the capital cost for this industry off the books of the integrators and onto the lands (and homes) of these farmers, thereby shifting about half the risk of failure away from the company and onto the farmers.

In practice, the farmers take more than half the risk because the integrators also seek to shift all of the environmental responsibility of raising poultry onto the farmer. Using an "independent contract," the integrator contracts with the farmer to raise the poultry owned by the integrator from the time the integrator brings them to the farmer's chicken houses until the birds are picked up. The farmers feed them the feed owned and delivered by the integrator, along with the medications and additives (insecticides, antibiotics, arsenic, etc.), also owned and delivered by the integrator. The poultry are kept in houses designed by the integrator, but paid for by the farmer. Some have referred to these farmers as serfs on their own farms.[23]

The contract provides that the birds remain the property of the integrator as long as they are alive, but if they die, they are the sole property of the farmer. The contract provides that everything that goes into the birds (except water) is owned and controlled by the integrator, but that everything that comes out of the birds as litter and urine is the sole responsibility of the farmer.

"Integrator liability" refers to the shift of environmental responsibility, either shared or complete, from the farmer to the company. It is based on the premise that companies are responsible for more of the production costs. Until the unfair externalization of environmental responsibility is corrected, integrators are not paying their fair share for the environmental problems they cause. The Sierra Club believes that the "independent contractor" scheme is an illegal action by the companies, and there is some indication that the courts are beginning to agree.[24] In Kentucky, the Sierra Club has sued Tyson over air emissions from three farms, seeking a judicial determination that Tyson must take responsibility for the environmental problems caused by its birds.[25]

Kentucky governor Paul Patton and former Natural Resources and Environmental Protection Secretary Bickford were both proponents of integrator liability. They promulgated emergency and permanent regulations requiring limited integrator liability for swine facilities in 1997, and in 1999 new regulations were implemented for all CAFOs. These regulations were challenged by the Farm Bureau and several other farm organizations.[26] The Farm Bureau, not surprisingly (given its allegiance to industrial farming practices), claimed that integrator liability was "unfair—like holding the car owner responsible when she leaves her car with Jiffy Lube, and Jiffy Lube illegally disposes of the waste oil." The Farm Bureau's comparison of a farm to a car brings to mind Wendell Berry's criticism of the industrial economy for treating living things as if they were machines: "To treat creatures as machines is an error with large practical implications."[27]

A series of court decisions and legislative actions have left the Kentucky administrative regulations in limbo.[28] However, the governor's support for integrator liability has served to protect Kentucky from the threatened invasion of swine CAFOs that loomed in 1997.

The proposed CAFO regulations published by the EPA in December 2000 include a clear recognition that those who own the birds and the feed are responsible for the waste created by the birds.[29] Unfortunately, the EPA under President George W. Bush ignored the language explaining the proposed rule, and did not require "co-permitting." In spite of this setback from the administration, the Sierra Club will continue to seek to hold integrators fully liable for the environmental impacts caused by their animals.

Reduced to the fewest possible number of words, my message is as follow: *Kentucky is not for sale!* Kentucky will no longer be available to nonresident empire builders who see our natural resources, including our people, as commodities. Kentucky will be open and welcome to all residents who will farm for or provide other commerce for their own community. *But Kentucky is not for sale!*

Words matter. Words that lead to action matter more than words that do not lead to action. I have described one type of action—a campaign—to oppose industrial agriculture and to support the type of agriculture that will eventually replace industrial agriculture. I offer it as a model for those who want to develop their own campaigns, and I offer an invitation to all to join us in our campaign. The well-planned campaign thoughtfully applies our resources and our strengths to the fight while carefully shielding our weaknesses.

I am confident about the future of agrarianism because the pillars on which agrarianism rest are sound and enduring, and the structures that today prop up industrial agriculture are at odds with the world. Still, we need to recognize that the agrarian cause involves us in a serious fight: a fight for our lives. We—or our children or their children—will win. If we can agree that we are in this fight, and we lift our arms, legs, and voices, and we have confidence and courage, we will win earlier. If we lack confidence and courage, and we despair and doubt, we—or our children or their children—will win, but it will take much longer and be much more painful.

I move forward with faith and hope rather than despair. Wendell Berry reminds us that the Gospel instructs us to have hope, and that hope is a virtue, ranking close behind love and faith. Hope is clearly better than despair. If you have to choose between these two, choose hope. The way to move from despair to hope is to take action, and the most effective way to take action is to become organized. However, one should go beyond hope. Gathering people together to change the world solely on the basis of hope implicitly admits doubt. I invite you to take organized action, and so help current despair grow into faith and confidence.

Notes

1. Wendell Berry. *The Unsettling of America* (San Francisco: Sierra Club Books, 1977, Preface).

2. "But the arch destroyers are the shepherds, with their flocks of hooved locusts, sweeping over the ground like a fire, and trampling down every rod that escapes the plow as completely as if the whole plain were a cottage garden plot without a fence." John Muir, *The Mountains of California* (San Francisco. Sierra Club Books, 1989), 266.

The Sierra Club's conflicts with ranchers in the western United States have been bitter, especially in areas of public land used by ranchers. At times the Sierra Club has taken, or has appeared to take, an absolutist position, opposing all grazing on all public lands under any circumstances. However, the official Sierra Club policy on grazing on public lands is not an absolute policy, although it puts a heavy burden on ranchers to demonstrate little or no environmental harm.

3. Sierra Ag policy: http://www.sierraclub.org/policy/conservation/agriculture.asp

Sierra Grazing policy: http://www.sierraclub.org/policy/conservation/grazing.asp

4. Clean Water Act, section 502 (14); 33 U.S.C.A.section 1362(14). "The term "point source" means any discernible, confined and discrete conveyance, including but not limited to any pipe, ditch, channel, tunnel, conduit, well, discrete fissure, container, rolling stock, concentrated animal feeding operation, or vessel or other floating craft, from which pollutants are or may be discharged. This term does not include agricultural stormwater discharges and return flows from irrigated agriculture." That act then defines "discharge of a pollutant" as "any addition of any pollutant to navigable waters from any point source." CWA

section 502(12); 33 U.S.C.A section 1362 (12). The act outlaws the discharge of any pollutant by any person except where that person has complied with the permitting requirements of the Act. CWA section 301; 33 U.S.C.A. Section 1311.

5. CWA section 402; 33 U.S.C.A. Section 1342 describes the "National pollutant discharge elimination system" permits that apply to all point sources of pollution.

6. Pollution from sources other than "point sources" is described as "nonpoint sources of pollution," addressed at CWA section 319; 33 U.S.C.A. section 1329, without separate regulatory requirements.

7. CWA section 303(d); 33 U.S.C.A. Section 1313, requires each state to prepare a list of impaired water bodies, where existing discharge limits are not sufficient to remove the causes of impairment, and to determine the pollution loading limit for that water that will remedy the impairment; the total maximum daily load (TMDL) is then intended to be allocated among the point sources and the nonpoint sources of pollution to bring the impaired water body back into compliance. See *Pronsolino v. Nastri*, 291 F.3d 1123 (9th Cir. 2002).

8. Quoted in *CARE v. Henry Bosma Dairy*, 65 F. Supp.2d 1129 (E.D. Wa. 1999).

9. 40 CFR section 122.23, Concentrated animal feeding operations (applicable to State NPDES programs, see section 123.25); 40 CFR Part 412—Feedlots Point Source Category.

10. See Kendall Thu et al., "Water Quality" (paper published in *Understanding the Impacts of Large-Scale Swine Production, Proceedings from an Interdisciplinary Scientific Workshop*, held June 29–30, 1995, Des Moines, Iowa); and Michael Mallin, "Impacts of Industrial Animal Production on Rivers and Estuaries," *American Scientist* 88, no.1 (Jan.-Feb. 2000): 26–37.

11. On March 6, 1997, Buckman Hog Farms #1 applied to the NREPC for permit for a CAFO to raise swine for Carroll Foods, which permit was issued by NREPC on March 21, 1997, without any notice to the public or to adjoining property owners. Fortunately, one adjoining property owner, John Wilson, declined Carroll Foods' offer to sell and instead decided to contest the permit. On June 30, 1998, the permit was revoked by agreement to settle the permit challenge. See *John Wilson and Sierra Club v. NREPC and Kenneth Buckman*, NREPC DOW File No. DOW-23471–042.

12. See *Concerned Area Residents for the Environment v. Southview Farm*, 34 F.3d 114 (2nd Cir. 1994), where the U.S. Court of Appeals found a twenty-two hundred head dairy herd was a "concentrated animal feeding operation," holding that the spray trucks that applied wet manure were point sources, and that the run off from the spray fields was a discharge, so that the facility was illegally discharging without a Clean Water Act permit.

See also *NRDC v. Costle*, 568 F.2d 1369 (D.C. Cir. 1977): " In sum, we conclude that the existence of uniform national effluent limitations is not a necessary precondition for incorporating into the NPDES program pollution from agricultural, silviculture and storm water runoff point sources." *Id.*, at page 1379.

See also *Carr v. Alta Verde Industries, Inc.*, 931 F.2d 1055 (5th Cir. 1991).

See the recent case of *Community Association for Restoration of the Environment (CARE) v. Henry Bosma Dairy et al.*, No. 01–35261 (9th Cir. September 16, 2002)

13. On October 30, 1989, Natural Resources Defense Council and Public Citizen filed suit against EPA alleging among other things that EPA failed to comply with Clean Water Act section 304(m). *Natural Resources Defense Council, Inc. et al. v. Reilly*, No. 89–2980 (RCL) (D.D.C.). Pursuant to the agreed upon settlement, the EPA administrator would issue proposed rules for CAFOs by December 15, 2000, and final rules by December 15, 2002.

14. See *Federal Register* 66, no. 9, Friday, January 12, 2001, pages 2960 et seq., and *Federal Register* 68, no. 29, Wednesday, February 12, 2003, pages 7175 et seq.

15. The National Environmental Policy Act (NEPA), 42 U.S.C.A. Section 4321 et seq., requires generally that all action of the federal government with the potential to have a significant impact upon the environment must first be subject to the preparation of an environmental impact statement (EIS).

16. See *Sierra Club v. USDA*, 234 F.3d 1269, 2000 WL 1679473 (6th Cir. November 2, 2000); See *Sierra Club, et al. v. Cagle-Keystone Foods et al.*, NREPC DOW File No. DOW-23778–043. The "hay farm" had seven sinkholes in the vicinity of the proposed waste treatment lagoons and in the vicinity of the area where the wastewater would be land-applied; the facility was approved without any treatment for pathogens based upon the claim that the "hay farm" would "attenuate" the pathogens.

17. Cagle's-Keystone letter of December 8, 2000, to city of Albany, on file with author.

18. Cumberland County, Kentucky, Fiscal Court Ordinance No. 199899–03, was challenged by James and Betty Upchurch on appeal to the Kentucky Court of Appeals after an unsuccessful challenge in the Cumberland Circuit Court. See *James P. Upchurch et al. v. Cumberland County Fiscal Court*, Court of Appeals, Case No. 2000–CA-002607. On January 31, 2003, a divided Court of Appeals issued an opinion affirming the Summary Judgment of the Cumberland Circuit Court. On February 18, 2003, James and Betty Upchurch filed a motion for discretionary review seeking Kentucky Supreme Court review; the matter is pending.

19. http://www.sierraclub.org/factoryfarms/

20. http://www.sierraclub.org/factoryfarms/rapsheets/; http://www.sierraclub.org/roadtrip/lowplainsdrifter/

21. http://www.sierraclub.org/watersentinels/

22. http://www.sierraclub.org/factoryfarms/antibiotics/

23. http://www.state.ia.us/government/ag/index.html. See Donald Stull, "To-

bacco Barns and Chicken Houses: Agricultural Transformation in Western Kentucky," *Human Organization* 59, no. 2 (2000): 157.

24. See the recent case from the Alabama Supreme Court, *Tyson Foods, Inc. v. Stevens*, 783 So. 2d 804 (Ala. 2000).

25. *Sierra Club, Inc., v. Tyson Foods, Inc.,* U.S.D.C., W.D. Ky., Civil Action No. 4:02 CV-073–MR.

26. See the Kentucky swine regulations at 401 KAR 5:009 and the Kentucky concentrated animal feeding operations regulation at 401 KAR 5:072E (emergency regulations) and 401 KAR 5:072, replaced with 401 KAR 5:074E and 401 KAR 5:074. See *Kentucky Farm Bureau et al. v. Commonwealth of Kentucky, NREPC, et al.*, Franklin Circuit Court, Div. I, Civil Action No. 00–CI-00706.

27. Wendell Berry, "For Love of the Land," *Sierra Magazine* (May/June 2002): 54.

28. See *Hon. Paul E. Patton v. Robert Sherman (LRC) et al.*, Franklin Circuit Court, Div. II, Civil Action No. 01–CI-00660.

29. See *Federal Register* 66, no. 9, January 12, 2001, page 3024: "EPA believes that ownership of the animals establishes an ownership interest in the pollutant generating activity at the CAFO that is sufficient to hold the owner of the animals responsible for the discharge of pollutants from the CAFO."

14

Private Property Rights in Land

An Agrarian View

Eric T. Freyfogle

An agrarian worldview is one that respects the land and its mysteries, that honors healthy, enduring bonds between people and place, and that situates land users within a social order that links past to future. Is there a particular understanding of private landownership that arises out of this perspective, or that might best sustain it? If there is, is it possible for us to implement this understanding, to shift from a mix of private rights and responsibilities strongly slanted toward development and industry to one that protects agrarians and their settled ways of life?

These questions ought to command more attention by agrarians than they do. The fact that they do not is probably due mostly to two factors: agrarians are themselves committed to private property, and they assume that the rights and responsibilities of ownership are essentially unchanging and unchangeable. The commitment to private property is well grounded in policy; it can stay. It is the latter assumption that needs rethinking.

Before the coming of the industrial revolution, cheap fossil fuels, and laissez-faire individualism, property law in America protected agrarian ways of life. That property system gave way in the nineteenth century to an alternative regime that fostered industrialization while sanctioning the abuse of lands and waters. Today's sick lands and communities display the scars of this largely forgotten legal shift. Now, with the advent of conservation, the pendulum has swung part way back as ecologically informed defenders of rights of ownership are paying attention again to nature and to ways of life linked to fertile land. It needs to keep swinging, much further, if agrarianism is to regain ground.

No agrarian needs reminding that the jobs for today are many. Yet few tasks offer more promise than the crafting of a new understanding of what it means to own land. For agrarians to labor within the present system is to abide by rules that strongly favor the opposition.

The prospects for such a new understanding of ownership, and what (in brief outline) it might look like, are the subjects of this essay.[1]

New York, 1805: While New York was still a British colony, one Palmer erected a sawmill on land that he owned along the Hudson River in Saratoga County.[2] Because the river was a public highway Palmer could not block it entirely, but as owner of shoreland he could construct a dam that extended well into the river. Once in place, his dam raised a head of water to supply power for his new mill. It also provided a convenient way to collect and store logs being floated to the mill.

Years later, a competitor constructed a similar dam and mill a mere two hundred yards upstream. The new mill, Palmer complained in the lawsuit that he soon filed, injured his business and interfered with his property rights. His principal injury was the greater expense he incurred in getting logs to his mill. Logs could reach the mill only if laborers carefully navigated them around the new upstream dam. The costs of doing that were sizeable, and hundreds of logs were washed down the river during the maneuver.

Palmer no doubt went to court with a high degree of confidence in his case, for the law was clearly on his side. As a riparian landowner he possessed the right to make use of any waterway that flowed alongside his land. The upstream landowner had the same right, but the water-use right that the law protected—the "natural flow" rule—was sharply limited. "Every man has a right to have the advantage of a flow of water, in his own land, without diminution or alteration," one of the justices explained. In operation, this strongly worded rule gave landowners only the right to use water in ways that left downstream owners unaffected. Except for household and subsistence uses, a riparian landowner could not alter the quantity, quality, or timing of a water flow to the detriment of others. In this instance, the upstream landowner was disrupting the Hudson's natural flow in ways that cost Palmer money.

By the time the drama ended Palmer had lost his case. In time, historians would view the contentious, three-to-two ruling as a sign that property rights in America were on the move.[3]

Writing in support of the upstream defendants, Justice Brockholst Livingston admitted that the "no harm" rule was "a familiar maxim" of property law. *Sic utere tuo ut alienum laedas*, courts had long phrased it: "use your own so as to cause no harm." The natural flow rule was merely a specific application of this protective doctrine. Applied rigorously, *sic utere tuo* meant that no land use could cause harm to other landowners or to the public at large, even if it was otherwise reasonable. The rule was meritorious in Livingston's view, and yet its application to the facts of Palmer's dispute seemed neither wise nor fair. Mills and dams such as the defendants' benefited the public, Livingston observed. The law should encourage such structures, not suppress them. The upstream landowners, moreover, had the same property right to use the river as Palmer did. To hold them liable—or even worse, to insist that they remove their dam—would deprive them of their own right to use what they owned.

Packed into Livingston's assessment of public policy was a line of reasoning that was surprising given the protection that the law had long afforded to a landowner's right to remain undisturbed when using his land. It was vital in resolving the dispute, Livingston asserted, to consider "the public, whose advantage is always to be regarded." In the past, the public interest had been equated with protecting landowners such as Palmer from being disturbed. No longer so, Livingston announced, or at least the protection of quiet enjoyment no longer overrode all competing considerations. The public good was also served by "competition and rivalry" among landowners. Hence, the right of individuals to "the free and undisturbed enjoyment of their property"—the right that Palmer was asserting—had to be weighed against "the public benefits which must frequently redound" from new land uses that incidentally harmed others.

To reconcile these competing policies Livingston proposed a significant shift in the law of waterway use, and by extension in private property rights generally. The upstream defendant should be allowed to continue using his dam, despite the *sic utere tuo* and natural flow rules that banned all harm, so long as the damage done to Palmer was not "manifest and serious." If the harm was less than that, Livingston ruled, Palmer simply had to put up with it.

It was a signal victory for industrial land users, anxious to use their lands intensively.

South Carolina, 1818. Astride his horse and in search of deer one Singleton entered into the unenclosed and unimproved land owned by M'Conico.[4] M'Conico ordered him to leave, Singleton refused, and M'Conico sued for trespass. A local jury ruled in favor of the wandering hunter, and the landowner appealed to the South Carolina Supreme Court. The rule of law that should govern the case, M'Conico asserted, was that a trespass occurred whenever a hunter failed to depart private land when ordered to do so, even if the land was unenclosed and unimproved.

In a five-to-one ruling the Court disagreed: "Until the bringing of this action, the right to hunt on unenclosed and uncultivated lands has never been disputed, and it is well known that it has been universally exercised from the first settlement of the country up to the present; and the time has been, when, in all probability, obedient as our ancestors were to the law of the country, a civil war would have been the consequence of such an attempt, even by the legislature, to enforce a restraint on this privilege." "The forest was regarded as a common" into which hunters had the right to enter at their pleasure, Justice Johnson explained. Forests served as a source of food for many of the state's citizens. In addition, public hunting on unenclosed lands allowed citizens to learn "the dexterous use and consequent certainty of firearms." Thus trained, citizens could ably serve as militia men, thereby allowing the state to avoid the costs and dangers of a large standing army. Open hunting served the common good.

Given Singleton's right to hunt on unenclosed lands, it seemed obvious, Johnson concluded, "that the dissent or disapprobation of the owner cannot deprive him of it; for I am sure it never entered the mind of any man, that a right which the law gives, can be defeated at the mere will and caprice of any individual."

For another day at least, the landowner's desire to exclude would remain less important than the rights of citizens to draw sustenance from the land.

Pennsylvania, 1886: In 1868 Mrs. Sanderson purchased land in the city of Scranton, Pennsylvania, near where Meadow Brook emptied into the Lackawanna River.[5] As a court later explained, "the existence of the stream, the purity of its water, and its utility for domestic and other purposes . . . was a leading inducement to the purchase." By 1870 Mrs. Sanderson had erected a house and built a dam across the stream to supply herself with fresh water, fish, and ice.

Upstream on Meadow Brook stood a sixteen-hundred-acre mining operation owned by the Pennsylvania Coal Company. Beginning about the time that Mrs. Sanderson bought her land, the coal company opened coal seams and sank mining shafts and tunnels. To keep the shafts dry, the company installed "powerful engines" to pump water to the surface, where it flowed into Meadow Brook through an artificial watercourse. The pumping not only materially increased the water flow in the stream, it also degraded the brook's water quality because the mine water contained natural impurities. Within a few years, the ill effects of the pumping had become manifest. The water in Mrs. Sanderson's pond had become "totally unfit for domestic use," the fish in the pond were dead, and pipes in the house were corroded. The hydraulic system that conveyed water into the house had been "rendered totally worthless," and in 1875 was abandoned. Mrs. Sanderson filed suit against the coal company, seeking to recover monetary damages for her injuries.

As it related the facts the Pennsylvania Supreme Court took pains to describe the mining company's valuable operations, which were conducted, it explained, in the normal manner of coal mines of the day and without negligence or malice. The excess mine water and its contaminants were all natural products of the land. Coal mining, moreover, was the natural and perhaps only possible use of the company's land, given its physical features. As for the water flowing into Mrs. Sanderson's pond, gravity was the agent that it took it there, without purposeful diversion by the mining company. Given these facts, the court asked rhetorically, why should the mining company be liable for the pollution?

Several facts weighed heavily upon the court as it pondered the dispute. Foremost was the importance of coal mining to Pennsylvania's economy. On her side, Mrs. Sanderson was only one person and the coal company could easily have covered her out-of-pocket losses. But if she could obtain damages as a riparian landowner for the harm done her, so too could all other landowners who were injured. With enough landowners lined up, hat in hand, the coal company's liability could mount up. Property rights, moreover, were perpetual, and it was "impossible to foresee what other modes of enjoyment" existing riparian owners or their successors in ownership might undertake in the future "or to estimate the extent of damages to which the continued pollution of the stream might proceed." Even more troubling than the danger of monetary liability for actual harm was the possibility that riparians such as Mrs. Sanderson would ask the court for punitive damages or for an injunction to halt all mining operations.

As the court saw things, an interpretation of nuisance law that allowed Mrs. Sanderson to recover money could in time bring the state's coal industry to an end. If the coal industry failed, the public in turn would suffer greatly. The court was not inclined to let it happen. In any event Mrs. Sanderson's injuries were not all that severe, or so the court callously asserted; they were a matter of "mere personal inconvenience," nothing more. "To encourage the development of the great natural resources of a country trifling inconveniences to particular persons must sometimes give way to the necessities of a great community."

The court turned next to the property rights that the coal company itself held.

> It may be stated, as a general proposition, that every man has the right to the natural use and enjoyment of his own property; and if, while lawfully in such use and enjoyment, without negligence or malice on his part, an unavoidable loss occurs to his neighbor, it is *damnum*

> *absque injuria* [injury for which there is no legal remedy]; for the rightful use of one's own land may cause damage to another, without any legal wrong. Mining in the ordinary form is the natural user of coal lands. They are, for the most part, unfit for any other use.

This, then, was the crux: the coal company itself had property rights, and on the facts of the case it was merely exercising those rights. "[E]very man is entitled to the ordinary and natural use and enjoyment of his property." So long as a landowner avoided negligence and malice, a land use was permissible, even when it severely disrupted neighbors.

For the time being, in Pennsylvania's coal company, the industrial view of property held the high ground.

Wisconsin, 1972. In 1961 Ronald and Kathryn Just purchased 36.4 acres of land along the south shore of Lake Noquebay in Marinette County.[6] Over the next half dozen years they severed and sold approximately three-quarters of their original tract in parcels. In 1967 the county enacted a shoreland protection ordinance that banned the deposit of fill material into designated wetlands without a permit. All of the land retained by the Justs was covered by the ordinance. Six months after the ordinance took effect, Ronald Just hauled in 1,040 cubic yards of sand and without a permit filled an area of twelve thousand square feet. When Marinette County objected the Justs filed suit, asking the court to declare the shoreland protection law unconstitutional because it took their property rights in violation of the "just compensation" clause.

As the court sized up the dispute, it posed "a conflict between the public interest in stopping the despoilation of natural resources, which our citizens until recently have taken as inevitable and for granted, and an owner's asserted right to use his property as he wishes." The effect of the county ordinance, the court observed, was not "to secure a benefit for the public" by compelling the Justs to leave their property unaltered; instead, it was "to prevent a harm from the change in the natural character" of the Justs' property. To gauge how severely the law curtailed the Justs' property rights one first had to decide what property rights they possessed in their wetlands. If the Justs had no right to fill them in, then the law took nothing from them. "Is the ownership of a parcel of land so absolute that man can change its nature to suit any of his purposes?" the court asked rhetorically. Assuredly not, it responded, at least not when public officials viewed the change as harmful to the public good: "An owner of land has no absolute and unlimited right to change the essential natural character of his land so as to use it for a purpose for which it was unsuited in its natural state and which injures the rights of others. The exercise of the police power in zoning must be reasonable and we think it is not an unreasonable exercise of that power to prevent harm to public rights by limiting the use of private property to its natural uses." Landowners possessed the right to use their lands "for natural and indigenous uses," much as earlier courts had declared. But such uses had to be "consistent with the nature of the land," which meant, in the case of wetlands, uses that were consistent with their continued retention as wetlands, not uses that required filling or draining. Because the county ordinance merely limited the Justs to uses consistent with the land's natural character, it worked no material hardship.

It was a startling, suggestive vision of ownership, with a forgotten past and an uncertain future.

Iowa, 1998. In 1993 the owners of 960 acres of farmland in Kossuth County, Iowa, petitioned the county board of supervisors to designate their land a protected "agricultural area" under the laws of the state of Iowa.[7] Their aim in seeking the designation was not to ward off development pressures, which were apparently modest. Instead, it was to gain the right as owners to engage in any agricultural activity that they wanted without fear that their enterprise could legally be deemed a nuisance. The landowners apparently sought the legal designation in order to set up the latest technology in meat production: the concentrated animal feeding operation (CAFO), which harbors in one place thousands or tens of thousands of animals along with their attendant odors, wastes, and pests. So long as the petitioning landowners complied with the specific rules governing CAFOs, an "agricultural area" designation would insulate them from the danger of a nuisance suit filed by neighbors. After balking at first, the county board voted by a narrow margin to grant the landowners' request.

Living nearby were Clarence and Caroline Bormann and Leonard and Cecelia McGuire. Fearing that the planning operation would disrupt their settled life, they filed suit in April 1995 to challenge the county's decision. The effect of it, they asserted, was to strip them of their legal rights as landowners to halt any activity that significantly and unreasonably interfered with the quiet use of their lands. So severe was this curtailment of their rights, they claimed, that it amounted to an unlawful taking of their property within the meaning of the federal and Iowa constitutions. Either the government should pay them for the taking or the "agricultural area" designation should be struck down.

Three years later, the Supreme Court of Iowa resolved the dispute, agreeing in full with the Bormanns' and McGuires' allegations. The county designation would indeed have stripped them of a valuable property right when it gave the owners of the 960 acres license to commit a nuisance. The "agricultural area" designation made the CAFO immune to a suit based on nuisance and therefore improperly interfered with the Bormanns' and McGuires' rights.

Had Mrs. Sanderson been alive, the ruling would have pleased her.

History. Among the useful lessons to draw from these scenes is the important truth that property law shifts over time, in terms of the rights and responsibilities of landownership. Statutes and ordinances regularly prescribe new land-use rules; judicial decisions gradually modify the underlying common law. The legitimacy of such legal shifts is easy to see. When the people are sovereign the law properly reflects their circumstances, needs, and aspirations. As these change, so too should the law.

Also evident in these scenes is how the right to own land at any given time is a mixture of the right to use it and the right to complain when neighbors interfere with one's use. The law can favor one or the other of these rights strongly, or it can attempt to balance the two evenly. It cannot, though, emphasize both, since a right to use land intensively necessarily comes at the expense of a neighbor's right to complain about disruptions.

Until the early years of the nineteenth century property law allowed landowners to engage only in "natural uses" of their lands and waters—usages that did not disrupt their neighbors' activities. *Sic utere tuo ut alienum laedas* was the guiding rule, which in practice barred all forms of land-use harm. The right to complain against interferences was high; the right to use land intensively was low. As the century progressed and America's enthusiasm for industrialization swelled and frothed, legal rules were rewritten. Not overtly, of course; that was not the way courts did their work, then or now. The law continued to tell landowners that they had to use their lands in ways that caused no harm. But this once-strict rule was now greatly qualified, as the Pennsylvania court in *Sanderson* so vividly illustrated. A landowner could not impose harm on others—unless, that is, he used his land in a way that benefited the economy and employed practices that were common to the industry, in which case he could get away with destruction.

The right to use land intensively had risen markedly over the century; the right to complain about interferences had correspondingly fallen. As one court put it, explaining and justifying the legal development, "the law is made for the times."[8] If the times favored intensive development, property law should not and would not bar the way.

So things stood at the beginning of the twentieth century. Not uniformly by any means, but by fits and starts, courts in the new century reconsidered their strong tilt toward industrial land uses. Legislatures and local zoning boards did so even more openly. Cases such as *Just* would remain on the fringe, in terms of their rejection of late-nineteenth-century values and understandings. But courts everywhere were becoming less tolerant of obvious harms, both to settled neighbors and sometimes even to the land itself.

The stories of shifting definitions of land-use harm, and of the varying ways that the rights of neighbors have fit together, are not the only useful tales embedded in the legal history of private property. Also woven into it is a vital, community-centered strand of legal thought: the idea that individuals are not the only holders of legal rights. Chief Justice Roger Taney distilled this reasoning in a prominent antebellum decision upholding government's power to override private rights when public needs demanded it: "While the rights of private property are sacredly guarded, we must not forget that the community also have rights, and that the happiness and well being of every citizen depends on their faithful preservation."[9] Related to this intellectual strand was the equally well-grounded doctrine that private property rights are necessarily constrained by the larger communal order. Justice Lemuel Shaw of Massachusetts set forth this understanding in an influential ruling from the 1850s:

> We think it is a settled principle, growing out of the nature of well ordered civil society, that every holder of property, however absolute and unqualified may be his title, holds it under the implied liability that his use of it may be so regulated, that it shall not be injurious to the equal enjoyment of others having an equal right to the enjoyment of their property, nor injurious to the rights of the community. All property in this commonwealth . . . is derived directly or indirectly from the government, and held subject to those general regulations, which are necessary to the common good and general welfare.[10]

To construct an agrarian understanding of private ownership, one need not go outside America's rich legal tradition to find good raw materials.

By 1906, perhaps 6 million people had read Henry George's *Progress and Poverty*, first published in 1879. With over one hundred editions in print it was an unlikely bestseller, a five-hundred-page volume on land ownership, political economy, and the causes of wealth. "Peddled in railway coaches and by candy 'butchers' along with the paperback joke books and thrillers of the day," historian Willard Hurst recorded, the book responded to "some pervasive, deep-felt need to probe and grasp for more understanding of cause and effect in social relations."[11] George himself became known as "the prophet of San Francisco." Among those captivated by George's thinking was writer Hamlin Garland, who wrote of his awakening in his autobiography:

> Up to this moment I had never read any book or essay in which our land system had been questioned. I had been raised in the belief that this was the best of all nations in the best of all possible worlds, in the happiest of all ages. . . .

> Now as I read this book, my mind following step by step the author's advance upon the citadel of privilege, I was forced to admit that his main thesis was right. Unrestricted individual ownership of the earth I acknowledged to be wrong and I caught some of the radiant plenty of George's ideal Commonwealth. The trumpet call of the closing pages filled me with a desire to battle for the right.[12]

From his home in California George journeyed east to New York City in 1869, where he was struck by the juxtaposition of vast wealth and extreme poverty.[13] As he probed this injustice, he found it rooted in America's land policy, which in turn was linked to larger economic questions. Paradoxically, America's growing wealth was bringing increased poverty along with it. Poverty in George's day was excused by many observers as an inevitable product of rising populations, just as conservative economist Thomas Malthus had said. Not so, George asserted passionately. Poverty had little or nothing to do with population growth. It arose because the country's lands and resources were all privately owned, and the poor had no choice but to work for their owners. As populations rose and competition for resources increased, more and more of the nation's wealth went to property owners. Something was wrong, not in nature, but in human institutions. Those who constructed buildings deserved a return on them; that was not the issue. It was the return on nature itself that bothered George, the return on vacant land, farmland, and land beneath tenement houses, factories, and stores.

George's lengthy analysis of the situation was complex and detailed, yet at its center stood a simple idea. Land values escalated as cities and economies expanded. Part of that rise came from improvements that landowners made, but raw nature itself was also increasing in value. Landowners were getting rich for doing nothing. As prices rose, rents increased. As rents rose, the landless paid greater portions of their income to live, and they took home in wages a falling percentage of what their labors created. The producer's right to the fruits of his labor was being frustrated by the landowning class.

George questioned the fairness of the entire idea that people could own nature individually. Nature was a gift to humankind in common, just as John Locke and others had said. But George was practical enough to realize that common ownership would not work in a complex world. Individuals had to have the right to control land individually. What would work, he believed, and what would be adequate to protect the public's interest, was for the public in some way to lay claim to the land values that the public itself had created. Rent from an office building could be divided between return on the building and return on the land, with the land portion going to the common fund. The same rule would apply to all lands, urban and rural. So considerable would be this source of public revenue by George's estimate that all other taxes could be removed. Government could live off a single tax.

George's logic seemed so powerful that adherents lined up by the thousand to carry his message. Single Tax clubs arose; Single Tax political rallies were held. Support for the idea swelled, despite the predictable resistance of landowners.

For many readers over the years George's analysis has seemed most useful, not when applied to urban office buildings and to rent paid on underlying land, but when applied to undeveloped land on the fringe. When development harms the public good, why should it be permitted? And when development is barred, why should the public be required to purchase "development rights"? The community itself has created the value of those rights: Why cannot the community simply curtail them?

Philosophy. Private property is an institution that at once expands individual liberty and contracts it. The secure ownership of things allows a person to flourish in ways not possible in a world that admits no private rights. A farm family, for instance, is free to grow and deepen its connection to the land when it holds reliable private rights in it. At the same time, the ownership of land by one person diminishes the activities that other people can undertake, most visibly when no-trespassing signs restrict where they can roam. In a world in which all land is privately owned, a landless person is severely constrained. To enforce these limits on the liberties of non-owners, the law allows owners to call upon police to arrest those who disregard their rights. Thus, private property and state power are inextricably linked. Like other forms of state power that restrict liberty, private property needs adequate justification if it is going to be legitimate.[14]

One common justification for private property derives from the inherent fairness of allowing a laborer who created a thing of value to own what she created. This reasoning holds merit, though it does require refinement to deal with certain weaknesses and limitations. In a world of scarcity, however, this labor theory (as it is termed) does not justify the private ownership of land, water flows, wildlife, and other parts of nature, which no human created. Rights in nature need justifying by some other line of reasoning. The only sound alternative is a justification linked to the welfare of the people collectively. Land ownership is good (and hence legitimate) under this justification because of the good consequences that arise when a society has it. This is a powerful justification for private property and yet it is at the same time limited because, under it, no private right is legitimate unless its recognition contributes to the overall good; to the good not just of the person who owns the thing, or of owners as a class generally, but of all citizens, owners and non-owners alike. It is not adequate merely to assert that landownership generally brings good to society and hence is legitimate: each landowner right (for instance, to build an industrial hog facility or to spray toxic chemicals) needs independent justification.

Such a utilitarian justification of private property can lead to the recognition of extensive private rights in all manner of things, when the public interest is served by them. But it can also lead to restrictions on such rights when and as the public interest evolves. Indeed, revisions in the legal terms of ownership may be necessary to avoid allowing owners to act in ways that are harmful to the public good. In the nineteenth century, lawmakers took away from landowners some of their rights to seek redress when their quiet enjoyment was disrupted by others; Mrs. Sanderson of Scranton was only one of the century's many losers. In the twentieth century, much legal change has occurred in the opposite direction, limiting the rights of owners to use their lands intensively while increasing their rights to complain about harmful acts by neighbors, near and far.

To dwell seriously on this utilitarian justification is to see why private property is properly understood as a creature of public law, rather than as some sort of natural or divinely created individual right. Property rights are defined and limited by law, mostly state law. When the law changes, they too change. Positivist reasoning of this type was familiar to America's founders and it gained broad currency in the nineteenth century. Benjamin Franklin was one of its strongest proponents during the Revolutionary Era. Private property was a "creature of society," Franklin confidently asserted, not some gift from God or nature. As such, it was "subject to the calls of that society, whenever its necessities shall require it, even to its last farthing."[15] Save for those few items "absolutely necessary" for subsistence—"the savage's temporary cabin, his matchcoat, and other little acquisitions"—all property was the product of "public convention," with the public holding full power to limit "the quantity and uses of it."[16]

Given this ongoing need to justify private rights by reference to their positive contributions to society's good, what basis is there to allow landowners to degrade nature in detrimental ways? Why should owners of water flows, for instance, have the right to use this vital resource in ways that cause environmental degradation?

Then there is Henry George's analysis, applied to modern circumstances. If development would harm the common good, why should landowners have the right to undertake it? And when development is banned, why should taxpayers be obliged to compensate owners, given the community's role in creating the development value?

Economics. To make sense of the economics of private land use, the place to begin is by recognizing that what happens on one land parcel inevitably has ripple effects that spread beyond property boundaries, ecologically, aesthetically, and socially. Not isolation but interconnection among land parcels is the norm in nature and in human communities. The implications of interconnection are several, chiefly that (1) a given land use, good or bad, necessarily affects other land parcels near and far; (2) neighbors and the surrounding community have legitimate interests in how private lands are used; and (3) many conservation goals (e.g., protecting wildlife populations or maintaining healthy rivers) are achievable only when landowners act in concert.

Writing in the 1930s, conservation theorist Aldo Leopold of the

University of Wisconsin paid particular attention to the economics of private land use and conservation.[17] Sound, conservative land use paid dividends overall, Leopold believed, at least in the long run, but the dividends were largely ones that landowners acting alone could not capture. Benefits spread to the entire community of which the landowner was only a part. When all landowners conserved, each might gain, just as everyone suffered when a single landowner failed to conserve. But conservation by an isolated owner rarely made economic sense.

For Leopold, these economic realities posed a challenge worthy of careful research. Repeatedly he would propose it as a topic: "the formulation of mechanisms for protecting the public interest in private land."[18] Existing institutions simply did not attend to the matter: "The present legal and economic structure, having been evolved on a more resistant terrain (Europe) and before the machine age, contains no suitable ready-made mechanisms for protecting the public interest in private land. It evolved at a time when the public had no interest in land except to help tame it."[19] Leopold worked hard on the challenge himself, identifying the tools available and assessing the relative merits of each. Economic incentives, education, legal restraints, boycotts, social ostracism, community-based conservation measures—Leopold would consider them all, only to find in time that none would do the trick, not alone at least and not in the combinations in which they had been used.

The central need, Leopold concluded, was for communities in some manner to insist that landowners act correctly. "Private land is only a stock certificate in a common biota," he would tell his wildlife ecology students. "Private land-use must recognize an obligation to community welfare. No other motive has enough coverage to suffice."[20]

Conservation writers since Leopold have added to his economic understandings by probing the special challenges that confront conservation work at the landscape scale. Dividing a natural commons into private shares can help avoid problems of resource overuse, just as Garret Hardin so famously observed, but it can make more difficult conservation efforts that require coordinated action at large spatial scales. The more a landscape is fragmented among individual owners, the harder it is to develop a coordinated plan: the "tragedy of the commons" that Hardin described is matched at the landscape level by the "tragedy of fragmentation."[21] Also more clear since Leopold's day is the profound influence of the pressures owners face when competing in national and international markets. Pressures to reduce costs can compel landowners to abuse land or risk bankruptcy. Free trade is a powerful engine of land degradation.

So what elements and attributes might characterize a private property system that took agrarian values seriously? How might property law better honor the local community and encourage if not compel owners to take better long-term care?

Land-use harm. For starters, an agrarian property system would likely define land-use harm in ways that protected sensitive land uses and that curtailed the powers of owners to degrade what they own. It would reduce the rights of owners to use the land intensively while increasing their rights to be protected against activities that drag them down.

Landowners have never had the right to engage in harmful uses, and the concept of harm has long been a flexible one, subject to ongoing redefinition. In an agrarian system, the do-no-harm, *sic utere tuo* doctrine would likely return to something close to the meaning it had late in the eighteenth century, before industrialization gained the upper hand.

Tailoring rights to the land. In an agrarian property regime those who wanted to engage in particular land uses would typically need to find lands that are well suited for them naturally, as well as economically and socially. The reasoning of *Just v. Marinette County* here is instructive: the land itself should play a role in setting limits on what landowners can do. Some land alteration would of course be allowed, but it would be much less than ordinarily permitted today. In an agrarian world, the landowner would be expected to fit the vision to the land, not the land to the vision.

Development rights. Agrarian culture emphasizes a land parcel's use value—its attractiveness as a place to build a home and draw sustenance from the land—rather than its potential market value when diverted to non-agrarian uses. In an agrarian property system, this form of land use would likely enjoy special protection.

The institution of private property is flexible enough to restrict a landowner's right to develop to those settings in which the community as well as the landowner benefits from it. An agrarian property regime would likely include such a restriction, in one form or another. Because land parcels vary so much, and because surrounding land uses and circumstances also vary, such an approach would doubtless produce a legal regime in which development rights differed widely among landowners, inevitably raising the specter of unequal treatment. Why should some owners get to develop and not others? A good question, deserving of an answer. Fortunately, good answers are available, or they are when restrictions are sensibly tied to real differences among the land parcels. In addition, legal tools are available to make such a regime more fair.

Who can own and how much? Episodically, American lawmakers have imposed limits on who could own land in a particular place. Agrarians today should consider if it would be wise for the law to limit both who could own farmland and how much they could own. Other countries have such rules, and they can work. Might farm ownership, for instance, be limited to those actively engaged in farming? Might farm sizes be limited so as to weed out land-use practices that inevitably degrade? As for laws limiting farm size, many precedents exist in American history (typically aimed at restricting land speculation by outsiders). A less heavy-handed way to protect agrarian families and practices would be to limit benefits under farm-support programs and under real-estate tax valuation schemes: only lands used and occupied by a farming owner, for instance, or farms up to some maximum size, might qualify for support payments or be entitled to lower farmland tax valuations.

Common lands. The United States has only weak traditions of lands that community members own collectively and manage as community resources. Somewhat more widespread, at least until recently, are traditions that allow the public to use unenclosed private lands for hunting and foraging so long as they do not disturb an owner's activities. In strictly economic terms, there is probably little reason today to try to revive and expand the notion of the town commons. But in social and educational terms the benefits of doing so could be great. Lands that local residents use together can become meeting places that foster social cohesion. Community farms and forests can provide good chances for young people to become attached to land and to learn something about subsistence land uses, as historian Brian Donahue has ably illustrated.[22] Rural areas might also provide attractive recreational opportunities, aiding local economies in the process, if far more acres were open to hikers, birders, and other recreational users. Practical matters aside, the idea of nature as a commons, or as commonwealth, could help reorient the ways people think about land and their relationship to it.

Acting together. One of the strengths and weaknesses of agrarian culture is its deep-seated sense of individualism and individual responsibility. The benefits of responsible individualism require no comment. The costs of it, however, do, for they are high. Leopold's comments about the economics of conservation provide one place to enter this issue. One could also enter it from the core of property law, with its longstanding focus on promoting the good of the community as such.

In the thinking of free-market advocates, the good of the community is best calculated by simply summing up the desires of the individuals who compose it. The community as such, that is—considered apart from its individual members—has no independent good that needs worrying about. This reasoning is distinctly faulty (as agrarians well know), but it is widely accepted nonetheless, in part because it leads so directly to wide freedoms for people to act however they please. Agrarians ought to be among those resisting this reasoning, and they should also resist its potent corollary—that the market should be given free rein to operate because it is the most efficient way to supply people with the goods and services that they want. In truth, the market is nowhere near so capable. It works well at giving people what they want as isolated individuals; the things that they can consume or enjoy themselves, without sharing with others. It does poorly at providing goods that benefit all people living in an area, whether they pay for it or not. Most conservation benefits are of the latter type. The market increases private wealth; the common wealth it destroys.

As Aldo Leopold explained, conservation is often economically sound in the long run when everyone engages in it, but the economics become adverse when landowners go it alone, some conserving, others not. Such a voluntary system is unfair to those who conserve—they incur the costs of conservation while having to share the benefits with those who do not conserve. Beyond that, there is the severe problem of unfair competition in the production of commodities and goods for the market. Why should a landowner who uses land conservatively, keeping waterways healthy, for instance, and leaving room for wildlife, be forced to compete in the market with owners who cut costs by harming the land and future generations?

For these reasons and others, conservation faces a steep uphill climb when landowners who share a landscape do not come together to address common problems. If particular land-use practices are dragging down the community and its long-term prospects, then the community ought to take action. If a landscape's carrying capacity is overburdened and cutbacks of some type need to made, collective action is essential.

More than they have, agrarians need to work together.

In thinking about this issue, agrarians might do well to pay attention to the work of other longstanding critics of the modern capitalist order. Marxist thought, for all its deficiencies, has done a better job than agrarian thought ever has (at least since the 1930s) in probing the realities of economic and political power. The capitalist market, Marxists have shown, is a potent force that degrades laborers, farmers, small communities, and the land. As isolated individuals, people can do little to fight back; they either fit within the market, drop out, or to varying degrees get crushed. To respond effectively people need to act in concert. Agrarians need to band together, not along the class lines that Marx recommended, but by forming coalitions with other people who also care about communities, land, and the economic ability of people to live peaceful lives.

In the end, it is hard to imagine how agrarian values and practices can thrive so long as agrarians remain gripped by an ethic of individualism. In the case of private property, concerted action by agrarians and others could lead to a new understanding of ownership that fosters rather than inhibits a thriving agrarian order. But no amount of fussing and fuming by isolated agrarians is going to bring it about. Without organized pressure, the industrial order will rule on.

NOTES

1. A more expansive consideration of the issues taken up here is offered in Eric T. Freyfogle, *The Land We Share: Private Property and the Common Good* (Washington, D.C.: Island Press/Shearwater Books, 2003).

2. Palmer v. Mulligan, 3 Cai. R. 307, 2 Am. Dec. 270 (New York Supreme Court, 1805). The facts of the dispute are recounted in the various opinions by court members.

3. A leading inquiry is Morton J. Horwitz, *The Transformation of American Law 1780–1860* (Cambridge: Harvard University Press, 1977). "*Palmer v. Mulligan* represents the beginning of a gradual acceptance of the idea that the ownership of property implies above all the right to develop that property for business purposes" (37).

4. M'Conico v. Singleton, 9 S.C.L. (2 Mill) 244 (South Carolina Supreme Court, 1818).

5. Pennsylvania Coal Co. v. Sanderson, 113 Pa. 126, 6 A. 453 (Pennsylvania Supreme Court, 1886).

6. Just v. Marinette County, 201 N.W.2d 761 (Wisconsin Supreme Court, 1972).

7. Bormann v. Board of Supervisors of Kossuth County, 584 N.W.2d 309 (Iowa Supreme Court, 1998).

8. Lexington & Ohio Rail Road v. Applegate, 8 Dana 289 (Kentucky Supreme Court, 1839).

9. Charles River Bridge v. Warren Bridge, 36 U.S., 11 Pet. 420, 548 (United States Supreme Court, 1837).

10. Commonwealth v. Alger, 7 Cush. 53 (Massachusetts Supreme Judicial Court, 1851).

11. James Willard Hurst, *Law and the Conditions of Freedom in the Nineteenth-Century United States* (Madison: University of Wisconsin Press, 1956), 74.

12. Hamlin Garland, *A Son of the Middle Border* (1917; reprint, New York: Penguin Books, 1995), 252.

13. Information on George and the writing of *Progress and Poverty* is taken from the book itself and from the introduction to the twenty-fifth-anniversary edition by Henry George Jr. in 1905 (reprinted in the fiftieth-anniversary Modern Library edition published by Random House, 1930).

14. A useful introduction to the various justifications for private property is Lawrence C. Becker, *Property Rights: Philosophic Foundations* (London: Routledge & Kegan Paul, 1977).

15. Benjamin Franklin, *Queries and Remarks Respecting Alterations in the Constitution of Pennsylvania,* vol. 10 of *The Writings of Benjamin Franklin,* ed. A. Smyth (1907), 54, 59.

16. Benjamin Franklin, *The Political Thought of Benjamin Franklin,* ed. Ralph Ketcham (Indianapolis: Bobbs-Merrill, 1965), 358.

17. His major writings on the subject include the essays "Conservation Economics" (1934) and "The Conservation Ethic" (1933), both reprinted in Susan L.

Flader and J. Baird Callicott, *The River of the Mother of God and Other Essays by Aldo Leopold* (Madison: University of Wisconsin Press, 1991), 181, 193.

18. Aldo Leopold, *Land Pathology* (1935), in Flader and Callicott, *The River of the Mother of God,* 212, 215.

19. Ibid., 214.

20. Aldo Leopold, Motives for Conservation (undated manuscript, Aldo Leopold Archives 10-6, box 14, University of Wisconsin, Madison).

21. See Eric T. Freyfogle, "The Tragedy of Fragmentation," *Environmental Law Reporter* (Envt. L. Inst.), 32 (2002): 11321.

22. Brian Donahue, *Reclaiming the Commons: Community Farms and Forests in a New England Town* (New Haven: Yale U.P., 1999).

15

Going to Work

Wendell Berry

I. To live, we must go to work.

II. To work, we must work in a place.

III. Work affects everything in the place where it is done: the nature of the place itself and what is naturally there, the local ecosystem and watershed, the local landscape and its productivity, the local human neighborhood, the local memory.

IV. Much modern work is done in academic or professional or industrial or electronic enclosures. The work is thus enclosed in order to achieve a space of separation between the workers and the effects of their work. The enclosure permits the workers to think that they are working nowhere or anywhere—in their careers or specialties, perhaps, or in "cyberspace."

V. Nevertheless, their work will have a precise and practical influence, first on the place where it is being done, and then on every place where its products are used, on every place where its attitude toward its products is felt, on every place to which its by-products are carried.

VI. There is, in short, no way to escape the problems of effect and influence.

VII. The responsibility of the worker is to confront these problems and deal justly with them. How is this possible?

VIII. It is possible only if the worker knows and accepts the reality of the context of the work. The problems of effect and influence are inescapable because, whether acknowledged or not, work always has a context. Work must "take place." It takes place in a neighborhood and in a commonwealth.

IX. What, therefore, must we have in mind when we go to work? If we go to work with the aim of working well, we must have a lot in mind. What must we know? We can establish the curriculum by a series of questions:

X. *Who are we*? That is, who are we as we approach the work in its inevitable place? Where are we from, and what did we learn there, and (if we have left) why did we leave? What have we learned, starting perhaps with the influences that surrounded us before birth? What have we learned in school? More important, what have we learned *out* of school? What knowledge have we mastered? What skills? What tools? What affections, loyalties, and allegiances have we formed? What do we bring to the work?

XI. *Where are we*? What is the place in which we are preparing to do our work? What has happened here in geologic time? What has happened here in human time? What is the nature, what is the *genius*, of this place? What, if we weren't here, would nature be doing here? What will the nature of the place permit us to do here without exhausting either the place itself or the birthright of those who will come later? What, even, might nature help us to do here? Under what conditions, imposed both by the genius of the place and the genius of our arts, might our work here be healthful and beautiful?

XII. What do we have, in this place and in ourselves, that is good? What do we need? What do we want? How much of the good that is here, that we now have, are we willing to give up in the effort to have further goods that we need, that we think we need, or that we want?

XIII. And so our curriculum of questions, revealing what we have in mind, brings us to the crisis of the modern world. Partly this crisis is a confusion between needs and wants. Partly it is a crisis of rationality.

XIV. The confusion between needs and wants is, of course, fundamental. And let us make no mistake: This is an *educated* confusion. Modern education systems have pretty consciously encouraged young people to think of their wants as needs. And the schools have increasingly advertised education as a way of getting what one wants; so that now, by a fairly logical progression, schools are understood by politicians and school bureaucrats merely as servants of the "the economy." And by "the economy" they do not mean local households, livelihoods, and landscapes; they mean the corporate economy.

XV. But the idea that schools can have everything to do with the corporate economy and nothing to do with the health of their local watersheds and ecosystems and communities is a falsehood that has now run its course. It is a falsehood and nothing else.

XVI. What actually *do* we need? We might say that, at a minimum, we need food, clothing, and shelter. And, if we are wise, we might hasten to add that we don't want to live a minimal life; we would also count comfort, pleasure, health, and beauty as necessities. And then, with the realization that it may be possible by reducing our needs to reduce our humanity, we may want to say also that we will need to remember our history; we will need to preserve teachings and artifacts from the past; we will need leisure to study and contemplate these things; we will need towns or cities, places of economic and cultural exchange; we will need clean air to breathe, clean water to drink, wholesome food to eat, a healthful countryside, places in which we can know the natural world—and so on.

XVII. Well, now we see that in attempting to solve our problem we have run back into it. We have seen that in order to understand ourselves as fully human we have to define our necessities pretty broadly. How do we know when we have passed from needs to wants, from necessity to frivolity?

XVIII. That is an extremely difficult and troubling question, which is why it is also an extremely interesting question and one that we should not cease to ask. I can't answer it fully or confidently, but will only say in passing that our great modern error is the belief that we must invariably give up one thing in order to have another. It is possible, for instance, to find comfort, pleasure, and beauty in food, clothing, and shelter. It is possible to find pleasure and beauty and even "recreation" in work. It is possible to have farms that do not waste and poison the natural world. It is possible to have productive forests that are not treated as "crops." It is possible to have cities that are ecologically, economically, socially, culturally, and architecturally continuous with their landscapes. It is not invariably necessary to *travel* from a need to its satisfaction or from one satisfaction to another.

XIX. It is not invariably necessary to give up one good thing in order to have another. In our age of the world there is a kind of mind that is trying to be totally rational, which is in effect to say totally economic. This mind is now dominant. It is always telling us that the good things we have are really not as good as they seem, that they can seem good only to "backward people," and they certainly are not as good as the things we will have in the future, if only we will give up the things that seem good to us now. If a forest or a farm is destroyed to make a "housing development," and the "housing development" is then sacrificed to a factory or an airport, the rational mind wants us to believe that this course of changes is "progress," and it offers as proof the successive increases in the value or the profitability of the land. It shows us the "cost-benefit ratio." And here we arrive at the crisis of rationality. We have come to the point at which reason fails.

XX. Reason fails precisely in the inability of the cost-benefit ratio to include all the costs. We know that, however favorable may be the cost-benefit ratio, the progress from forest or farm to any sort of "development" degrades or destroys the integrity of the local ecosystem and watershed, and we know that it causes human heartbreak. This kind of accounting excludes all coherences except its own, and it excludes affection. The cost-benefit ratio is limited to what is handily quantifiable, namely money. The failure of reason comes to light in the recognition that things that cannot be quantified—the health of watersheds, the integrity of ecosystems, the wholeness of human hearts—ultimately affect the durability, availability, and affordability of necessary quantities. To think of landscapes merely in terms of economic value will in the long run reduce their economic value, not to mention the availability of such necessities as timber and food, clean water and clean air.

XXI. The mind makes itself totally rational in an effort to become totally comfortable, but at the risk of eventually becoming totally uncomfortable. The cost of subordinating all value to economic value will eventually be economic failure.

XXII. We are well acquainted with this mind of would-be total rationality, hell-bent on quantification. And we are increasingly well acquainted with its results in the ruin of culture and nature. And so the next in our curriculum of questions necessarily is this: Do we know of a different or better or saner kind of mind?

XXIII. I think we do. It is what I would call the affectionate or sympathetic mind. This mind is not irrational, but neither is it primarily rational. It is a mind less comfortable than the mind that aspires only to reason, and it is more difficult to define.

XXIV. It is defined, I think, in the parable of the lost sheep in the Gospels of Matthew and Luke, and in the Buddhist vow "Sentient beings are numberless, I vow to save them." The mind given over to reason would lose no time in demonstrating mathematically that it "makes no sense" to leave ninety-nine sheep perhaps in danger while you go to look for only one that is lost. And surely it makes even less "sense" to vow to "save" all sentient beings.

XXV. Obviously, to assent to such teachings as these we must change our minds. We must give up some part of our allegiance to reason and to quantification, and we must accept as our lot in life a perhaps irreducible discomfort. We have given affection and sympathy a priority over rationality. We have consented to the proposition that at least a part of what we have now, the part we have been given, is good, and we have assumed the responsibility of preserving the good that we have. We have assented implicitly to God's approval of His work on Creation's seventh day.

XXVI. To change one's mind in this way is to change the way one works. This changed way of working is new to us in our industrial age, but is old in the history of human making. And what is this way? How does this changed mind go about its work?

XVII. Such a mind, I think, is no longer satisfied with the conventional standards of industrialism: profitability and utility. Needing a more authentic, more comprehensive criterion it looks beyond those concerns, without necessarily giving them up. It tries to see the work and the product in context; it tries to derive its standards from that context. And once again it must proceed by way of questions: Is the worker diminished or in any other way abused by this work? What is the effect of the work upon the place, its ecosystem, its watershed, its atmosphere, its community?

What is the effect of the product upon its user, and upon the place where it is used?

XVIII. Work under the discipline of such questions might hope to give us, to name a few examples, forestry that would not destroy the forest, farming that would not destroy the land, houses that would be suited to their places in the landscape, products of all kinds that would neither exhaust their sources nor degrade their users.

XXIX. Obviously, there has come to be a radical disconnection between the arts and sciences and their ultimate context, which is always the natural or the given world. Why should this be?

XXX. I venture to think that it is a problem of perception, most particularly and directly in the sciences. The scientific need for predictability or replicability forces perception into abstraction. The "test plot," for example, is perceived, not as itself, but as a plot *representative* of all plots everywhere.

XXXI. Developers of technology, in somewhat the same way, are under commercial pressure for *general* applicability. The place where a new machine or chemical or technique is proved workable is assumed to be *representative* of all places where it might work.

XXXII. These processes in science and technology seem to be closely parallel in effect to the processes of centralization in economic and political power.

XXXIII. The result is that all landscapes, and the people and other creatures in them, are being manipulated for profit by people who can neither see them in their particularity nor care particularly about them.

XXXIV. The disciplines that are not directly involved in this manipulation nonetheless have consented to it. It is the problem of *all* the disciplines.

XXXV. It seems to me that the solution to this problem is not now foreseeable, but I believe it can come about only by widening the context of all intellectual work and of teaching—perhaps to the width of the local landscape.

XXXVI. To accept so wide a context, the disciplines would have to move away from strict or exclusive professionalism. This does not imply giving up professional competence or professional standards, which have their place, but professionalism as we now understand it has already shown itself to be inadequate to a wide context. To bring local landscapes within what Wes Jackson calls "the boundary of consideration," professional people of all sorts will have to feel the emotions and take the risks of amateurism. They will have to get out of their "fields," so to speak, and into the watershed, the ecosystem, and the community; and they will have to be actuated by affection.

XXXVII. In the sciences, I think the acceptance of the local landscape as context will end the era of scientific heroism. No one scientist or one team of scientists or one science-exploiting corporation can expect to "save the world," once the disciplines have accepted this context that is at once wide and local. The solutions then will have to be local, and there will have to be a myriad of them. The scientists, moreover, will have to suffer the responsibility of applying their knowledge at home, sharing the fate of the place where their knowledge is applied.

XXXVIII. Throughout these notes I have been assuming—as my reading and the work I have done have taught me to assume—that it is impossible for us humans to know in any complete or final way what we are doing.

XXXIX. Now I will explain this assumption in a different way, but one that leads to the same conclusion.

XL. Increasingly since the Renaissance, the building blocks of rational thought have been facts, pieces of data that can be proved or demonstrated or observed to be "true."

XLI. The assumption seems to be that the pursuit of truth in our time, as never in the times before, has become completely scientific and rational, so that now we not only possess more facts every day than we ever possessed before, but have only to continue to fill in the gaps between facts by the empirical processes of our science until we will know the ultimate and entire truth.

XLII. I do not believe this. I think it is a kind of folly to assume that new knowledge is necessarily truer than old knowledge, or that empirical truth is truer than non-empirical truth. But I also do not believe that factual truth is or ever can be sufficient truth, let alone ultimate truth.

XLIII. A fact, I assume, is not a thing, but is something known about a thing. The formula H_2O is a fact about water; it is not water. A person who had never seen water could not recognize it, much less recognize ice or steam, from knowing the formula. Recognition would require knowledge of many more facts. Water is water because it is the absolute sum of all the facts about itself, and it would be itself whether or not humans knew all the facts.

XLIV. The only true representation of a thing, we can say, is the thing itself. This is true also of a person. It is true of a place. It is true of the world and all its creatures. The only true picture of Reality is Reality itself.

XLV. In order to work, in order to live, we humans necessarily make what we might call pictures of our world, of our places, of ourselves and one another. But these pictures are artifacts, human-made. And we can make them only by selection, choosing some things to put in the picture, and leaving out all the rest.

XLVI. From the standpoint of the person, place, or thing itself, of Reality itself, it doesn't make any difference whether our pictures are factual or imagined, made by science or by art or by both. All of them literally are fictions—things made by humans, things never equal to the reality they are about and never assuredly even adequate to the reality they are about.

XLVII. Facts in isolation are false. The more isolated a fact or a set of facts is, the more false it is. A fact is true in the absolute sense only in association with *all* facts. This is why the departmentalization of knowledge in our colleges and universities is fundamentally wrong.

XLVIII. Because our pictures of realities, and of Reality, are necessarily incomplete, they are always to some degree false and misleading. If they become too selective, if they exclude too much (on the ground, for instance, of being "not factual"), if they are too biased, they become dangerous. They are constantly subject to correction—by new facts, of course, but also by experience, by intuition, and by faith.

XLVIII. We may say, then, that our sciences and arts owe a certain courtesy to Reality, and that this courtesy can be enacted only by humility, reverence, propriety of scale, and good workmanship.

Further Reading

Berry, Wendell. *The Art of the Commonplace: The Agrarian Essays of Wendell Berry.* Edited by Norman Wirzba. Washington, D.C.: Counterpoint, 2002. This collection gathers essays from the wide corpus of Berry's work (including *The Gift of Good Land, Home Economics, Another Turn of the Crank,* and *What Are People For?*), presenting agrarianism as a coherent challenge to industrial/technological culture.

———. *Farming: A Handbook.* New York: Harcourt Brace Jovanovich, 1970. A poetic rendition of agrarian life and its ideals.

———. *The Unsettling of America: Culture and Agriculture.* 3rd ed. San Francisco: Sierra Club Books, 1996. This book stands as the definitive contemporary statement of agrarian concerns and priorities.

Buell, Lawrence. *Writing for an Endangered World: Literature, Culture, and Environment in the U.S. and Beyond.* Cambridge: Harvard University Press, 2001. A leading ecological literary critic examines the place of the physical environment in the work of nineteenth- and twentieth-century writers.

Donahue, Brian. *Reclaiming the Commons: Community Farms and Forests in a New England Town.* New Haven: Yale University Press, 1999. A lively account of the development of Land's Sake community farm in Weston, Massachusetts, and an elaboration of its philosophical principles.

Freyfogle, Eric T., ed. *The New Agrarianism: Land, Culture, and the Community of Life.* Washington, D.C.: Island Press, 2001. Fifteen essays from many leading contemporary agrarians advocating cultural reform.

Hanson, Victor Davis. *The Land Was Everything: Letters from an American Farmer.* New York: Free Press, 2000. A classicist and farmer, in a series of letters patterned on J. Hector St. John de Crèvecoeur's well-known *Letters from an American Farmer* (1782), makes an impassioned plea for the recovery of virtues gained by working the land.

Jackson, Wes. *Becoming Native to This Place.* Washington, D.C.: Counterpoint, 1994. A contemporary statement of a new farming economy based on nature's principles, and focused on the revitalization of rural communities.

———. *New Roots for Agriculture.* Lincoln: University of Nebraska Press, 1980. A critique of industrial agriculture and a proposal for "natural systems agriculture" as being developed at the Land Institute.

Jacobs, Jane. *The Nature of Economies.* New York: Modern Library, 2000. The most recent book by one of our most renowned and pioneering urban theorists (*The Death and Life of Great American Cities,* 1961) on the link between culture and habitat, economies and natural processes.

Logsdon, Gene. *The Contrary Farmer.* White River Junction, Vt.: Chelsea Green Publishing, 1994. A spirited explication of the agrarian worldview, with perceptive criticism of contemporary culture.

———. *Living at Nature's Pace: Farming and the American Dream.* 2nd ed. White River Junction, Vt.: Chelsea Green Publishing, 2000. Essays on various agrarian themes.

———. *The Man Who Created Paradise: A Fable.* Athens: Ohio University Press, 1998. A short, hopeful, visionary statement of agrarianism in action restoring a strip-mined landscape.

Orr, David. *Earth in Mind: On Education, Environment, and the Human Prospect.* Washington, D.C.: Island Press, 1994. A powerful case for the restructuring of education in light of ecological realities.

Pollan, Michael. *Second Nature: A Gardener's Education.* New York: Dell Publishing, 1991. Though staged as a gardening book, this entertaining and wide-ranging work illuminates many agrarian and cultural concerns.

Redekop, Calvin, ed. *Creation and the Environment: An Anabaptist Perspective on a Sustainable World.* Baltimore: Johns Hopkins University Press, 2000. A collection of essays examining religious dimensions of land stewardship.

Savage, Scott, ed. *The Plain Reader: Essays on Making a Simple Life.* New York: Ballantine, 1998. Inspired by Amish, Quaker, and "plain folk" ways, this collection of essays considers the application of agrarian values to everyday life.

Schama, Simon. *Landscape and Memory.* New York: Vintage Books, 1995. Though not explicitly about agrarianism, this capacious book chronicles the place of landscapes in our long cultural memory.

Shi, David E. *The Simple Life: Plain Living and High Thinking in American Culture.* New York: Oxford University Press, 1985. A well-written history of American attempts to instantiate agrarian, among other, ideals.

Shiva, Vandana. *Monocultures of the Mind: Perspectives on Biodiversity and Biotechnology.* London: Zed Books, 1993. A vigorous critique of monocultures in agriculture and society and a defense of ecological and cultural diversity.

Smith, Kimberly K. *Wendell Berry and the Agrarian Tradition: A Common Grace.* Lawrence: University Press of Kansas, 2003. An extended analysis of Wendell Berry's work as it connects with and critiques American political and social thought.

Soule, Judith D., and Jon K. Piper. *Farming in Nature's Image: An Ecological Approach to Agriculture.* Washington, D.C.: Island Press, 1992. A detailed account of the failure of modern industrial agriculture and a defense of ecologically informed agriculture. Builds on Jackson's *New Roots for Agriculture.*

Thompson, Paul B., and Thomas C. Hilde, eds. *The Agrarian Roots of Pragmatism.*

Nashville: Vanderbilt University Press, 2000. Various essays develop agrarianism as a philosophical position within the American traditions of pragmatism and letters.

Twelve Southerners. *I'll Take My Stand: The South and the Agrarian Tradition.* Baton Rouge: Louisiana State University Press, 1962. Originally published in 1930, this manifesto by John Crow Ransom, Allen Tate, Robert Penn Warren, and others served as a rallying cry for those disenchanted with industrialism.

Vitek, William, and Wes Jackson, eds. *Rooted in the Land: Essays on Community and Place.* New Haven: Yale University Press, 1996. A diverse collection of essays expanding on several central agrarian themes.

Williams, Raymond. *The Country and the City.* New York: Oxford University Press, 1973. A classic examination of English literary treatments of rural/urban life.

Wirzba, Norman. *The Paradise of God: Renewing Religion in an Ecological Age.* New York: Oxford University Press, 2003. An application of agrarian principles to religious and cultural renewal.

Worster, Donald. *The Wealth of Nature: Environmental History and the Ecological Imagination.* New York: Oxford University Press, 1993. Essays by a leading environmental historian on agrarian and ecological themes.

Contributors

Wendell Berry is the recipient of several literary awards and author of numerous books, including most recently *Life is a Miracle* and *Jayber Crow.* He farms near Port Royal, Kentucky.

Herman E. Daly is currently a professor at the University of Maryland, School of Public Affairs. From 1988 to 1994 he was a senior economist in the environment department of the World Bank. He is the author of *Beyond Growth* and *Ecological Economics and the Ecology of Economics.*

Brian Donahue is an assistant professor and director of the environmental studies program at Brandeis University. He is the author of *Reclaiming the Commons* and *The Great Meadow: The Nature of Husbandry in Colonial Concord, Massachusetts.*

Eric T. Freyfogle is the author of *The Land We Share: Private Property and the Common Good* and editor of *The New Agrarianism: Land, Culture, and the Community of Life.* He teaches property, natural resources, and environmental law at the University of Illinois College of Law.

Hank Graddy has held several offices with the Sierra Club and currently chairs the Sierra Club CAFO/Clean Water Campaign Committee. He practices environmental, civil, and land-use law in Midway, Kentucky.

Wes Jackson is president of the Land Institute in Salina, Kansas. He is the author of *New Roots for Agriculture* and *Becoming Native to This Place.*

Frederick Kirschenmann is the director of the Leopold Center at Iowa State University and president of Kirschenmann Family Farms, an organic grain and livestock farm in North Dakota.

Benjamin J. Bruxvoort Lipscomb is an assistant professor of philosophy at Houghton College. He and his family live in Fillmore, New York.

Gene Logsdon is the author of numerous publications. His more recent books include *The Contrary Farmer* and *The Contrary Farmer's Invitation to Gardening*. He lives near Upper Sandusky, Ohio.

Benjamin Northrup was a trained carpenter before becoming a lecturer in the Yale School of Architecture. He currently practices traditional architecture in New Haven, Connecticut, with a special interest in urban design and town planning.

David W. Orr is professor and chair of the Environmental Studies program at Oberlin College and author most recently of *The Nature of Design*.

Vandana Shiva is the director of the Research Foundation for Science, Technology, and Natural Resource Policy in India. She is the author of many books, including *Water Wars* and *Stolen Harvest: The Hijacking of the Global Food Supply*.

Maurice Telleen edits *The Draft Horse Journal* in Waverly, Iowa. He is a member of the Prairie Writer's Circle.

Norman Wirzba chairs the philosophy department at Georgetown College, Kentucky, and is the author of *The Paradise of God: Renewing Religion in an Ecological Age* and editor of *The Art of the Commonplace: The Agrarian Essays of Wendell Berry*.

Susan Witt has served as the executive director of the E.F. Schumacher Society since its founding in 1980. Her essays have appeared in numerous journals and books, including *Orion Magazine* and *A Forest of Voices: Conversations in Ecology*.

Index